Pavlov's Dogs and Schrödinger's Cat

Pavlov's Dogs and Schrödinger's Cat

Scenes from the Living Laboratory

Rom Harré

OXFORD
UNIVERSITY PRESS

OXFORD
UNIVERSITY PRESS

Great Clarendon Street, Oxford OX2 6DP

Oxford University Press is a department of the University of Oxford.
It furthers the University's objective of excellence in research, scholarship,
and education by publishing worldwide in

Oxford New York

Auckland Cape Town Dar es Salaam Hong Kong Karachi
Kuala Lumpur Madrid Melbourne Mexico City Nairobi
New Delhi Shanghai Taipei Toronto

With offices in

Argentina Austria Brazil Chile Czech Republic France Greece
Guatemala Hungary Italy Japan Poland Portugal Singapore
South Korea Switzerland Thailand Turkey Ukraine Vietnam

Published in the United States
by Oxford University Press Inc., New York

First published 2009

British Library Cataloguing in Publication Data
Data available

Library of Congress Cataloging in Publication Data
Library of Congress Control Number: 2008942646

Typeset by SPI Publisher Services, Pondicherry, India
Printed in Great Britain
on acid-free paper by
CPI Antony Rowe, Chippenham, Wiltshire

ISBN 978-0-19-923856-9

1 3 5 7 9 10 8 6 4 2

In memory of Nero, mighty hunter and the best of cats

Table of Contents

List of Illustrations ix

Preface xi

1. Instruments and Apparatus: Tools for Experimenting 1

2. Detecting: Hyenas, Frogs, Zebra Fish, and Assorted Farm Animals 20

3. Measuring: The Bodies of Kings, Fossils, Lichen, and Pollen Grains 57

4. Extending an Established Domain: Dogs, Horses, Frogs, and Trees 89

5. Exploring a New Domain: A Bald Mouse, Two Chimpanzees, and Some People 123

6. Testing Hypotheses: More Dogs, Monkeys, and Ticks 147

7. Modelling Individuals: Yet More Dogs, A Few Mice, and One Human Stomach 176

8. Modelling Worlds: Voles, Peas, Fruit Flies, Finches, and Deep Oceanic Worms 202

9. **Practising Deception: Spring Wheat and Midwife Toads** 242

10. **Inventing Novel Beings: An Imaginary Cat and Virtual Life Forms** 269

Reflections: Sites for Moral Debates 297

End Notes 302

Index 313

List of Illustrations

1. The laboratory at the Royal Institution (Mary Evans Picture Library) 4
2. Experiments with frog's legs (Luigi Galvani, *De Viribis Electricitatis Motum Animalium* (Bologna, 1791)) 24
3. Montgolfier balloon (Getty Images) 51
4. Dog Laika in *Sputnik* (Mary Evans Picture Library/Interfoto Agentur) 54
5. Pollen grains (G. Erdtmann, *An Introduction to Pollen Analysis* (Waltham, MA, 1943)) 80
6. Pressure/volume relation for the blood of a horse (Stephen E. Hales, *Statical Essays*, vol. 2, *Haemastatics* (London, 1769)) 110

7. Leaf transpiration in and out of water — 115
(Stephen E. Hales, *Statical Essays*, vol. 1, *Vegetable Statics* (London, 1769))

8. Vacanti mouse — 126
(© BBC Picture Library)

9. The experimental setup. A subject in the obedience experiment at Yale — 140
(From the film *Obedience*, © 1968 by Stanley Milgram, © renewed 1993 by Alexandra Milgram and distributed by Penn State Media Sales)

10. Pavlov and dog — 153
(© Time & Life Pictures/Getty Images)

11. Surrogate mother monkey — 162
(Harlow Primate Laboratory, University of Wisconsin, Madison, USA)

12. *Helicobacter* microbes — 196
(Barry Marshall, *Helicobacter and PUD*)

13. Vole postures — 210
(J. R. Clarke, *The Aggressive Behaviour of the Vole* (Leiden, 1956))

14. Cells in a sample of cork — 216
(R. Hooke, *Micrographia* (London: Royal Society, 1665), 114)

15. Lysenko measures wheat — 253
(© Hulton-Deutsch Collection/Corbis)

16. Spin-up and/or spin-down — 283

17. Lineage of biomorphs — 293
(Richard Dawkins, *The Blind Watchmaker*, © 1996, 1987, 1986 Richard Dawkins. Used by permission of W. W. Norton & Company, Inc.)

Preface

No doubt the theories propounded by distinguished scientists and their popularizers are of great general interest. It can hardly be a bad thing to share some of the insights that thoughtful and informed people have had about the world and its inhabitants. However, at the root of any theory there lies a fertile ground of experimentation. Without this ground theorizing is idle. When one turns to study the manipulation and prodding of nature by experimenters one is struck by something as fascinating as anything theoreticians can offer. Experimentation is made possible by an array of pieces of equipment, instruments and apparatus, which have an interest all their own.

This book is a sequel to my *Great Scientific Experiments* study of 1981, in which twenty interesting and sometimes influential experiments were described and set in the lives of the ingenious people who performed them. In this book instruments and apparatus rather than the details of their use in experiments are the focus. But not just any instruments and apparatus—we will look at a selection of the many cases in which the equipment was organic, using living things or their parts and remains in much the same way as equipment of glass and metal has been used.

From the middle of the eighteenth century there have been voices raised against the use of living beings in experimental science. Alexander Pope, Bernard Shaw, and many others have found the exploitation of animal bodies for scientific purposes anathema. Less has been said about the uses of plants, though that too has a moral dimension. Careless and destructive treatment of plants can spread to our ways of treating other living beings. Attitudes to the use of living beings in scientific experimental programmes have ranged from complete laissez-faire, anything goes that leads to an increase in human knowledge, to the demands of some 'animal liberationists' for a total ban on all uses of organic beings in scientific research, except perhaps consenting humans. Most people, scientists included, probably hold to some middle position. What is lacking from much of the debate is a comprehensive view of how organic beings and their parts and remains have actually been used, together with some idea of the kind of people who have used them and for what scientific purposes. The aim of this book is to fill this gap. Standing back from history is possible only if one suspends one's own moral intuitions, marvelling sometimes at the characters and paradoxical mindsets of some of the dramatis personae of this story.

This is not a history. The studies are arranged according to the principles of philosophy of science. Nor is it a science text. Scientific topics of very different content and from very different times are put together in the chapters according to such categories as the logical principles of scientific thinking and particularly the use of models in science, both in experimenting and in theorizing.

To work across such a broad spectrum of topics and times as this book required has been greatly facilitated by the astonishing powers of the Internet. Searching through the millions of books in the Bodleian and Radcliffe libraries in Oxford and the Lauinger Library in Georgetown University would have been impossible without its help. Search engines reveal the books and articles germane to a topic in profusion. Then libraries bring forth these books and journals for first-hand examination. I particularly thank the staffs in the Oxford and Georgetown University libraries for their patience and expertise in finding the necessary texts for me with the utmost good humour. My particular thanks to Latha Menon, patient and sensitive editor, to whom I owe a great debt of gratitude for her invaluable help in putting this text into order.

Rom Harré
Linacre College, Oxford
Georgetown University, Washington, DC

Chapter 1
Instruments and Apparatus: Tools for Experimenting

Around 380 BC Aristotle took a clutch of fertilized eggs and by opening one every day for a month observed the daily stages in the development of the chick. Thus began the systematic study of embryology. Since then scientists have been using living beings as experimental apparatus in a vast array of studies in many different research fields. Sometimes there was an inorganic alternative. A coil of wire hanging by a thread can detect electric fields. So can a pair of frog's legs, though by the standards of early researchers, the legs were an instrument of greater sensitivity. Sometimes it seems that only living beings will do. Evolution is an organic process. There seems to be no non-living alternative to the use of animals and plants to study it. Gregor Mendel's use of pea plants and Rosemary and Peter Grant's choice of the finches on an isolated island as their experimental apparatus for modelling organic evolution seem inevitable. Yet, Richard Dawkins has led the way to the creation of virtual worlds, in which the inhabitants seem to evolve in a life-like way though in reality they are no more than images on a computer screen.

Adopted and adapted for laboratory uses animals and plants serve as equipment with which we experiment. In biology animals and plants are also the subjects of our experiments. This book is about the uses to which we put animals and plants as experimental equipment. Looked at this way animals and plants become devices we research *with* rather than something we research *on*. To see how living beings have been used in science we must begin by reflecting on the kinds of devices scientists use in gathering data and testing theories. Then we can follow the ways that living beings have been transformed into living forms of such equipment, into living instruments and apparatus.

Popular science books tend to present science in terms of ideas. Matt Ridley's admirable *Nature via Nurture* tells the story of biology and its relation to cultural and historical studies of human development as a conflict between the concepts that underlie and give meaning to rival theories.[1] Though Ridley describes the results of all sorts of experiments, he does not spend much time describing the equipment used. Stephen Hawking's *Short History of Time* covers the development of our *notions* of time and change in a cosmological setting.[2] However, most of the men and women who *do* science are at work in laboratories where they manipulate equipment to test hypotheses and to collect data.

Glass, metal, and ceramics as well as plants and animals have played a major role in the construction of laboratory equipment. Much of the routine aspects of experimental research have now been handed over to automatic devices, for example, performing and recording the results of chemical analyses. Though these devices work without human intervention, they incorporate instruments and apparatus in the traditional style. In most of the examples we will encounter in which living beings are involved as intrinsic parts of laboratory equipment we find few mechanized experimental procedures. For the most part, there are people working with real plants and animals and their parts and extracts. However, this is changing. In the last chapter we will look forward to an era in which computer-controlled experimentation and computer simulations will become more and more dominant in the armoury of the experimental scientist, even in the living laboratory.

Equipment for an experiment

Instruments and pieces of apparatus are the meeting point between the active experimenter and the potent stuff of the Universe. Not so long ago one would take for granted that when we open the door and peer into the 'lab' we will see something like the scene pictured in Figure 1, the laboratory of Michael Faraday. There will be lots of things made from glass and metal, wrought into shapes just right for their purposes.

Faraday's lab, like the school science laboratories of our youth, was filled with *inorganic* equipment. An apparatus for studying the action of hydrochloric acid on zinc could involve a *glass* flask, a cork with a thistle funnel and some *glass* tubing, running off to a small *metal* tank of water in which is an inverted *glass* gas jar. A device for detecting and measuring electric current might consist of various *metal* structures, such as a coil of wire and a magnet, suitably arranged. This is inorganic stuff. However, we would get a very limited view of the instrumentarium of the sciences,[3] the available instruments and pieces of apparatus, if we focused only on devices made of glass and metal. An important part of the instrumentarium is and has always been *organic*, apparatus and instruments constructed from living materials or materials that

Figure 1. The laboratory at the Royal Institution.

were once alive. Harry Harlow used live infant monkeys in his experiments on mothering. Roland Bischel used lichens to measure the age of rocks. Pollen grains became the basis by which Lennart von Post dated the layers of lake sediment, and extracts from lichens have served to detect the presence of acids and alkalis in solutions.

In this book we explore the various roles that the equipment of the living laboratory has been developed to fulfil. Just as there are inorganic devices, such as flasks, alembics, and Bunsen burners, so there are organic devices, like pairs of frog's legs, fossil molluscs, and pea plants. Just as there are inorganic instruments like the swinging coils of galvanometers so there are organic instruments like ozone-sensitive plants and the carbon monoxide-sensitive canaries.

As the story of each kind of organic device is introduced it will sometimes be matched by an account of an inorganic device, instrument, or apparatus. The point is not to suggest that the organic device could have been replaced by an inorganic analogue. Rather, it is to highlight the similarity of the logical thinking behind the construction of laboratory equipment of various kinds, whether made from living material or inorganic stuff.

INSTRUMENTS

Instruments and natural phenomena are related causally. Some state of the world is able to bring about a change or changes in an observable and sometimes quantifiable property of an instrument. There are all sorts of *detectors*, such as litmus paper sensitive to the presence of acids and alkalis, transgenic fish to the presence of pollutants, and pollen grain counts to climate change. There are also all sorts of *measuring* devices for ascertaining the relative quantities of material things, such as rulers, clocks, balances, and thermometers. Some of these are organic or derived from biological sources. Changes in the state of that to be measured must stand in some orderly relationship to observable changes brought about in the measuring device. The numbers and sequence of different species of fossil midges in the sediment on the floors of lakes can be used to determine the temperature of a lake at times past. The temperature at the time caused one species of gnat to predominate over others. Lichens grow slowly but steadily over hundreds of years. They can be used as simple and reliable clocks to measure the age of the rocks on which they are growing.

To reiterate: Instruments are related *causally* to the world.

APPARATUS

An important part of scientific research is to abstract one process, say a chemical reaction, from complex clusters of reactions in the chemistry of the real world, reproduce it in the laboratory, and study it in relative isolation in strictly controlled conditions. In a physics laboratory one might find a simple copper vessel containing a mixture of ice, salt, and water, a basic calorimeter. It can be used to reproduce the cooling of salt water below the temperature at which pure water freezes. This happens in the sea, but in a much more complex situation, with currents and landmasses exerting their particular influences on the formation of the winter ice which floats on the surface of the sea. The calorimeter provides knowledge of one part of the process of the freezing of the arctic sea in winter by drawing on the analogy between the water in the calorimeter and the ocean. The scientific value of experiments with the calorimeter depends on the balance of differences and similarities between the apparatus and the real-world phenomenon it represents.

The chemists of the eighteenth and nineteenth centuries were adept at constructing pieces of apparatus for isolating physical and chemical processes from the complex environments in which they occurred in nature. Lavoisier set up an apparatus that abstracted the role of atmospheric gases from all the other conditions involved in combustion. He isolated the process by which a metal was reduced to ash, and then followed the reactions by which the metal was eventually recovered. The power of this research depended on the fact that Lavoisier had abstracted the core processes of burning and smelting and reproduced them in his laboratory.

We *reason* from what happens in the laboratory to how things might be in nature. Because it happens thus and so in the laboratory, it is likely to happen thus and so in the world outside, all else being equal! The experiment contributes to our scientific knowledge because we believe that it reproduces an *instance* of a natural *regularity*, isolated from all the other processes going on in the real world. Stephen Hales managed to organize his experiments on the flow of sap so that only the transpiration of the leaves was left as a candidate to explain the movement of water upwards from the roots.

However, the reproduction of natural processes can go further than isolating just one aspect of a complex reality. Scientists attempt to create working models of the processes of nature in a more elaborate and realistic way. For his studies of the process of the formation of rainbows Theodoric of Freiberg set up racks of rows of spherical flasks filled with water. Each flask represented a raindrop, and the whole rack was a representation of the curtain of falling rain. Setting a lamp in the same relative position to the rack as the Sun would have been to a real rain shower, he was able to produce a working model of a complex natural phenomenon in his laboratory. Here he could do something impossible in nature—study the paths taken by rays of light of different colours through the water in flask/raindrop. Theodoric projected his discoveries about the path of light rays in spherical masses of water onto rain drops by the logical move of *reasoning* by analogy. This was an inorganic creation. There are plenty of logically similar creations in which the material stuff is organic. The

Grants were able to use the results of their observations on Daphne Major to support Darwinian evolutionary hypotheses of natural selection by drawing an analogy between the evolution of the bird population on the isolated world of their island and the whole of organic nature. Harry Harlow's wire frame with a teat was somewhat analogous to a monkey mother, and represented just one mothering behaviour abstracted from a very complex reality. A cloth-covered frame abstracted another aspect of monkey mothering to be studied in isolation from the real world. Was suckling more significant than comfort in the lives of infant monkeys? These abstract analogues enabled him to investigate the question without the interference of all sorts of other aspects of nurturance. Pieces of apparatus are used to realize simplified and sometimes isolated versions of natural processes, similar to and different from what they represent in relevant ways.

To reiterate: Apparatus is related *conceptually* to the world it helps us to understand.

THE CONTENTS OF THE INSTRUMENTARIUM

Looked at this way scientific equipment includes four main classes of devices. There are instruments for detecting the presence of something in the environment, such as the presence of carbon monoxide, by noticing the moment at which the canary ceases to sing. There are measuring instruments, for determining the quantitative values of properties of their targets, such as the age of a pollen grain as an indicator of the time at which lake-bottom sediments were laid down. There are pieces of apparatus in which natural processes can be studied in isolation, such as Mendel's garden plots from which he worked out the algebraic law of hybridized attributes of peas. There are pieces of apparatus which serve as models representing very complex processes and situations in nature in ways more convenient than studying what goes in the natural world itself. John Clarke established two colonies of animals to study their modes of life as a whole, though in the course of his researches he abstracted various key processes that influenced the fertility of the populations of his worlds.

R. E. Kohler describes his exposition of the work of geneticists with the fruit fly *Drosophila* thus:

> I will treat Drosophila as a scientific instrument, a piece of laboratory technology [though we will use 'apparatus' for this kind of device in this book]. Drosophila was not so much an object of study as a means for producing genetic knowledge. Standard organisms like Drosophila can be seen as systems of production, designed artifacts that have skills and procedures built into them through long use. Such instruments have many potential uses; the question is for what are they actually used, and how are such choices made?[4]

Here we have the 'living laboratory'.

Each laboratory has its characteristic instrumentarium, the actual equipment available to an experimenter. Depending on the generosity of the budget, the instrumentarium will consist not only of what is in the storeroom, but also what is to be found for purchase in the catalogues of instrument makers. In the great days of experimental science, in the time of Michael Faraday (1791–1867) for example, the laboratory was not only a place where instruments and pieces of apparatus were used, but also where instruments and apparatus were made. Science has always found a place for living beings in the instrumentarium. In Pavlov's laboratory the instrumentarium was almost exclusively populated by dogs as it was in Shumway's research institute. People filled the instrumentarium of Milgrim's laboratory.

Some working assumptions in the use of the equipment

INSTRUMENTS

To infer something about an object of interest from the state of pieces of equipment in use for making measurement or for detecting invisible and intangible things, substances and processes, in a laboratory or elsewhere, the scientist must believe that there has been a causal interaction between the world and the instrument. Generally it is assumed that the instrument is in a way detached from or outside the world it is being used to study. So an important underlying assumption on the use of instruments for detecting or measuring an attribute of an inhabitant of the natural world is that the equipment is affected by but does not affect that which it samples, or at least not much. Sometimes this assumption does not hold well and allowances for reciprocal effects must be made. Sometimes the thermometer requires so much heat to expand the mercury that the liquid being studied cools down substantially. Sometimes the experimental procedure kills the animal at the core of the apparatus, as when Pavlov snipped the vagus nerve to demonstrate nervic control of pancreatic secretions. Skilled experimenters know how to compensate for these exceptions. Usually the pressure in the car tyres is not significantly reduced by the amount needed to activate the tyre gauge. Ideally instruments are designed to minimize reciprocal causal reactions. What if a chimpanzee has its own agenda in interacting with those who are studying it? So far as I could see, Washoe, the chimp that was the main feature of early studies of the linguistic capacities of primates, was often manoeuvring the Gardners, the experimenters, into having fun at the same time as they were teaching her ASL. The games did not significantly interfere with the experiment, and indeed were made good use of.

The readings of instruments depend for their scientific value on causal relations that link states of the world to observable states of the instrument.

APPARATUS

Using a piece of laboratory equipment as a model for some natural process is related conceptually rather than causally to the rest of the material world, as I have emphasized. Of course an apparatus is part of that world. A map of the world is an object in the world, just as an authentic model village contains a model of itself. Again there is an isolation assumption involved. The model world contains as much of the real world as it needs to evolve its own forms of natural processes, so that though it is set aside from the world it represents whatever happens in the model as authentic. Sometimes the 'isolation' assumption is rudely disturbed, as when a predator from the larger world got into the isolated world where John Clarke's colonies of voles were living, and slaughtered many of them. Only rarely does the isolation assumption need be dealt with explicitly. For the most part apparatus as a model can be used as if it were detached from or outside the material world, as if it were a truly isolated system.

Models or analogue representations of aspects of the world under study are extremely common in the sciences. An analogue can be analysed in terms of likenesses and differences to that which it is analogous to. In the sciences model building is usually based on a process which is already well understood, and used to illuminate something which is not. The former is the *source* of the model, the latter its *subject*.

Darwin used the breeding of pigeons and farm animals as the source of his model of speciation, the process that he imagined had led to the diversity of species, calling it 'natural selection'.

The 'selection' idea, modelled on breeding practices, the farmer's selection of desirable features in the animals to breed from next season, was an untried analogue of what might be the natural process truly responsible for the origin of species. That mysterious process was the subject which Darwin's model was intended to represent. This idea of the trio 'Source'–'Model'–'Subject' will prove very useful when we analyse the role of living or once living beings as apparatus in scientific research.

Some equipment can be thought of as constructed for modelling a process, structure, or system which we want to understand in a simplified and manageable material setup. It often happens that the subject of the model cannot easily be manipulated or experimented on, but the model that represents it can be. For example, a culture of a bacterium thought to be responsible for an infectious disease can be grown on agar in a Petri dish. An infusion of a possible antibiotic is introduced and we watch for an effect on the bacteria of the culture. This is an *in vitro* version or model of a phenomenon which we have reason to believe occurs *in vivo*, when an antibiotic is given to a patient with a certain kind of infection. Apparatus like the simple Petri dish as a model of an infected human being represents only one aspect of a complex reality, the infective agent and the antibiotic. Modelling like this, creating a simplified version of real world complexities, is a very common scientific technique. We will see it exemplified in Pavlov's dogs, Mendel's peas, John Clarke's abandoned swimming pool stocked with voles, and so on.

Studying domesticated versions of natural things, systems, and processes supports quite strong inferences to the way things are in the wild, since the same *kind* of material systems and phenomena occur in the wild and in domestication. An apparatus, as a model, however much it is simplified and tidied up as a version of its source, is a piece of nature in the laboratory. Of course, the richness and authority of inferences from model to subject will depend on how relations of similarity and difference between apparatus made by technicians and real-world systems are weighted by the interests of the researcher in performing the experimental manipulations. These are the premises of reasoning by analogy. Since anything can in principle be analogous to anything else, the art of analogical reasoning requires a certain skill and discretion. Modelling is a scientific technique that requires a good deal of intuition and insight to be really effective. Nevertheless, it is the most powerful tool available to the experimenter from the instrumentarium.

Sometimes the modelling relation leads to hypotheses about what might have happened or what could be happening, processes that cannot be observed directly with any experimental apparatus or detector from the current instrumentarium.

In reflecting on the analogical power of a piece of apparatus we must ask whether it is just similarities in the way the apparatus and its subject *behave* that matters, what we could call the behavioural analogy, or are we ready to go further and to declare that we can also infer something about the *process*, hitherto unobserved that brings about that behaviour, the material analogy at the heart of a model. When the Grants followed natural selection on their little Galapagos Island, was it right to infer that the model could be generalized not only to the fossil record, but to the processes, in principle unobservable, that had brought about the changes in the forms of plants and animals in millions of generations hundreds of millions of years ago? The Grants observed changes in the populations of finch species and they also observed, at first hand, privation and sexual selection that Darwin had only imagined. This question will come up in several contexts, for example how far can one project the processes that make possible the life of micro-organisms in hot volcanic lakes in the Yellowstone Park of today back onto processes that must have underlain the transformation of the Earth's atmosphere from nitrogen and methane to its present mix of nitrogen and oxygen? Scientists studying extremeophiles have no hesitation in reasoning from similarities of behaviour to similarities in the nature of the primitive organisms, the archaea, that flourish in boiling pools to the nature of the earliest complex organisms and the life processes that made their existence possible.

Questions posed by the existence of the living laboratory

What plants and animals and parts of plants and animals have been used as apparatus? Pavlov used the whole of the dog's body as his apparatus, but Galvani adapted only the frog's legs as an electrical instrument. How have they been used? William Smith used an arrangement of fossils to determine the order in which geological strata had been laid down. Gregor Mendel used pea plants as the core of his research programme to study patterns of heredity. Who were the people who created living laboratories? Sandwiched between his mathematical studies and his feats as an architect Christopher Wren devised the equipment necessary to test ideas about blood transfusions. How successful have their projects been in scientific terms? Theodosius Dobzhansky used his colonies of drosophila to work out the fine details of inheritance, results which have become a permanent part of our knowledge of biology. By tracking the movement of the corpuscles in the blood vessels of the lung of a frog Stephen Hales was able to show that pulmonary circulation was many times more rapid than in other parts of the body. That result still stands. What has been the fate of the plants, animals, and plant and animal parts as a research programme has developed? Regretfully one must admit that most of the creatures that served as pieces of apparatus were 'sacrificed' or 'euthanized' after their role in the research programme had come to an end.[5]

People too are organisms. The use of the bodies of members of the species *Homo sapiens*, as experimental apparatus, *Man* considered as an animal, will also figure in these discussions. For the most part the lives of humans who have served as experimental equipment have been spared. Stanley Milgrim may have made some of his 'subjects' anxious but none of his experiments were fatal. Barry Marshall recovered from infecting himself with *helicobacter*. People have rarely been sacrificed or euthanized in the service of the growth of scientific knowledge. When this has happened it has provoked the strongest possible moral condemnation.

Most people have opinions about the living instrumentarium and the people who have used living beings in their experimental designs, or parts taken from living beings. In my experience onlookers from outside the world of science know little of the details of the uses that have been made of living beings as instruments and apparatus. Just as in every aspect of life, whatever judgements, moral or practical, we are inclined to make about the existence of the living laboratory can only benefit from a clear grasp of what sorts of organic equipment have been used and how animals and plants have been prepared for this special role. Glass must be skilfully blown to become a flask. A length of copper wire must be rolled up in a certain precise way to become an induction coil. So too living material must be shaped and transformed into the devices in the living instrumentarium. Frog's legs must be dissected out of the original animal in a very precise way with the crural nerves intact in order to detect electric currents; dogs must be reshaped by 'physiological surgery' so that the way that digestive secretions are prompted by stimuli can be studied; plots of peas must be planted, cultivated, and harvested and the different shapes and colours of the crop counted. How these transformations are achieved is surely very different in the inorganic and the organic domains. To make informed judgements we must come to know just exactly how Ivan Pavlov remade dogs; just what was required for the Grants to transform the bird population of a Galapagos Island into a huge living apparatus; what was needed to be able to read the age of a hyena's cave from the organic remains that littered the floor.

My project is to describe the work that people did in the living laboratory and at the same time to show, albeit often briefly, what sort of people they were and what unanswered questions pushed them on to carry through their researches. What was it that got the irascible and demanding Lennart von Post out of bed early on a freezing Swedish morning to go off, festooned with equipment, to try to recover sediment from the bottom of the local lakes? Incidents from the life stories of scientists cannot be properly presented without taking notice of the times in which these people lived, and the character of the society around them. Stephen Hales was a man of deep social conscience and was much troubled by the way the poorer people were treated in his times, the middle eighteenth century. He worked tirelessly to improve the conditions of life for sailors and prisoners in the stinking insanitary conditions of the day. He was instrumental in designing and installing ventilators in ships and prisons. But it was also he who 'stripped the skin from living frogs' to try to understand how muscular contractions were brought about!

This book is not about the experimental basis of the scientific study of life forms as such, the biological sciences. It is about the use of life forms in science in general for creating the apparatus needed for various experimental programmes, including the experimental study of biological topics. I make no moral judgements about the characters and motivations of the people I introduce and the work I describe. However, I hope that the organic material and its uses are described with the sort of accuracy and historical sensitivity on which informed moral judgements can be made. Christiaan Barnard drained the blood from a baboon, and replaced it with human blood, checking out the result the next morning on his way to the airport, finding the baboon eating an orange. Barnard came from a pious Afrikaans family and prayed before every operation he performed, including no doubt the experiment with the baboon. In the last section of this book I propose an 'ethical framework' of distinctions that strike me as relevant to moral debates around the material presented here. They might perhaps serve as a guide in thinking through some of the ethical issues that arise when we explore the instrumentarium of the living laboratory.

Commentary

As we have seen, the instrumentarium includes two broad kinds of laboratory equipment, instruments and apparatus.

In this book my aim is to describe the way that the bodies of plants and animals, including members of the species *Homo sapiens*, have been drawn into the scientific project in general. Sometimes it has been as instruments or parts of instruments, sometimes as apparatus or parts of apparatus. Each case will be presented as if the moral implications of these experimental programmes could be set aside until the story is told as fully as possible. At the same time, I have made no attempt to conceal the features of the scientific uses of plants and animals that demand moral scrutiny. Nor will I refer, except as an aside, to the good or ill any of these pieces of equipment in the hands of the scientists who designed, built, or commissioned them may have led to.

When the topic of 'animals in science' is raised in popular media, it is almost always in a moral context. When, if at all, does the welfare of human beings outweigh that of the animals used in research programmes which, it is hoped, will lead to some human good? Do animals have rights commensurable with the rights that people enjoy? And so on. In the last section of this book—'Reflections'—I lay out an ethical frame of distinctions which strike me as germane to any deep moral debate about the uses of animals in scientific research. I make a place in that frame for the possibility of raising moral issues about the uses of plants in science, not to be overlooked I believe.

Let me be quite clear at the outset of this study: I have no doubt at all that many animals enjoy a kind of subjectivity not unlike our own. In consequence, I have serious reservations about many projects in which animals have been involved, remembering that I too am an animal. I propose to leave my body, worn out from the study of philosophy and the instruction of generations of students, to the medical school, but they can have it only after I am dead! Short of a blanket prohibition on the uses of animals in science in any way, it is clear that difficult judgements must be made. The considerations are complex, and in many cases speculative. However, I believe that it is essential in weighing up the pros and cons of questions as to the rightness of the use of animals in pioneering surgical techniques for example, to have a very wide general picture of the many ways animals and plants have served as instruments and apparatus in living laboratories. So I have left the setting out of an ethical frame until the very end of this study.

Only when we have a clear idea of what has been done by whom and for what purpose can we take up the pressing moral questions that must arise. Virtual animals have no subjectivity. But how we treat imaginary animals may seep into the way we work with the instrumentarium of the living. There may be moral issues in the world of imaginary creatures and their role in thought experiments too.

Chapter 2
Detecting: Hyenas, Frogs, Zebra Fish, and Assorted Farm Animals

In this chapter we turn to the use of animals, birds, and plants, and their parts and extracts as instruments for detecting the presence of specific materials such as noxious gases, scientifically important states of the local environment such as electric charges as the intangible sources of electrical phenomena, and many other significant aspects of the natural world. We cannot perceive electric fields by their effects on our senses, perhaps a mild tingling at best, but we can create instruments that are visibly affected by them.

Instruments can also be used to scrutinize the observable properties of things more closely than the human senses. A human being can detect the presence of acids, alkalis, and other distinctive substances in solutions by taste or smell. This was how doctors used to detect sugar in the urine of diabetics. The drug squad still tastes the white powder to see if the plastic bags contain cocaine. In this case the tongue, the nose, and their neural connections are serving as an instrument. However, some acids and alkalis are corrosive, some are poisonous. When the question for the detectives is whether a half-drunk cup of tea contains cyanic acid we need something else than our tongues, though the smell of almonds is a rough guide. We need instruments to *detect* such things.

Instruments have been developed not only to reduce the risk to human observers of contact with dangerous materials, but also to overcome problems of distance and scale. How can we detect the presence of a microscopic bacterium in a blood sample taken from a sick person? It cannot be by taking a peek at the small red drop squeezed from the finger tip. A specimen is sent off to the medical laboratory to be cultured and identified by the use of an instrument designed to make the invisible visible, the microscope. Detecting things and events' distance in time is inseparable from their distance in space. The possibility and reliability of detection of distant events and conditions of distant things depends on the existence of extended causal chains, from the event in question to the instrument near at hand. Signal propagation takes time. So the event that is distant in space is also distant in time. At first glance one might think that the reaches of space and time are so vast that there would be no place in the instrumentarium for astronomical instruments based on living beings. However, for certain purposes, finding out whether extraterrestrial space is a safe environment for astronauts, animals have been used to try to detect hostile influences beyond the surface of the Earth.

An inorganic detector: the gold leaf electroscope, its origins and uses

Let us consider, for later comparison with the making of instruments from living beings and their parts, an example of an instrument for detecting an invisible, intangible something, made out of glass, metal, and ceramics. It makes use of no organic materials at all. The idea of a device to detect an insensible electric field must be credited to Jean Nollet (1700–70). His 'electroscope' of 1748 relied upon electrostatic forces. The version of the instrument still widely used in school science laboratories, the gold-leaf electroscope, was invented by Abraham Bennett (1749–99), curate of the Parish of Wirkworth in the county of Derby.[1] He describes a great many electrical and magnetic experiments in his *New Experiments in Electricity* (1789). Like Nollet's device, Bennett's electroscope detected the presence of an electric charge by the movement of a pair of gold leaves attached to a metal rod. The leaves were sheltered from draughts inside a glass flask. If, for example, a positively charged body was brought into contact with the metal lid of the flask both the leaves acquired a positive charge and so repelled each other. The stronger the charge the more the leaves diverged. If, however, the charged object had been brought close to the lid, but not in contact, it would induce an opposite or negative charge in the lid and so the leaves, picking up that charge, would both be negative. By the same principle that like charges repel one another, they would also diverge, one from the other.

But must the electroscope always be made of glass, metal, and ceramics?

Galvani and animal electricity

In the second half of the eighteenth century systematic studies of electrical phenomena became very popular. The names of two of the pioneers in the science of electricity have become part of our everyday vocabulary, Luigi Galvani and Alessandro Volta. At that time three kinds of electricity were recognized, going under a variety of names. Let us call them 'atmospheric electricity', manifested particularly in lightning; 'artificial electricity', produced by friction in electrical machines; and 'animal electricity', emitted by electric eels and seemingly involved in the workings of the nervous system. They were thought to be species of the same genus of phenomena because of the similarity in their effects. They could be made to cause sparks and they gave people shocks when they came into contact with them. Were they more closely related than species of the same genus? Did they have an identical nature?

Luigi Galvani's experiments, reported in detail in his *De Viribus Electricitate in Motu Muscularis* of 1791, involved the use of the hind legs of frogs as the essential core of experimental devices he developed to try to answer this question: what is the relation between 'animal electricity' and 'common electricity'? Just why he had dissected out pairs of legs with the crural nerves intact is not clear from the account in his book. A pair of such legs was apparently just lying around on the laboratory bench. He says, 'I dissected a frog and prepared it [as in Figure 2]. Having in mind other things, I placed the frog on the same table as an electric machine.'[2]

It had been known for several decades that muscular spasms could be produced by electrical stimulation. It had also been suggested that nerves were conductors of electricity, which it was supposed must be the true nature of 'animal spirits'. Felice Fontana (1730–1805) and others had noticed that the interior of neural fibres was inert and viscous, while the sheath was resinous. Perhaps the core was a conductor like a metal wire and the sheath an insulator. Galvani's contribution was his systematic experimental exploration of the hypothesis that the animal spirits were electrical, and that there was good reason to think that this was the same as the physicists' 'common or artificial electricity'.[3]

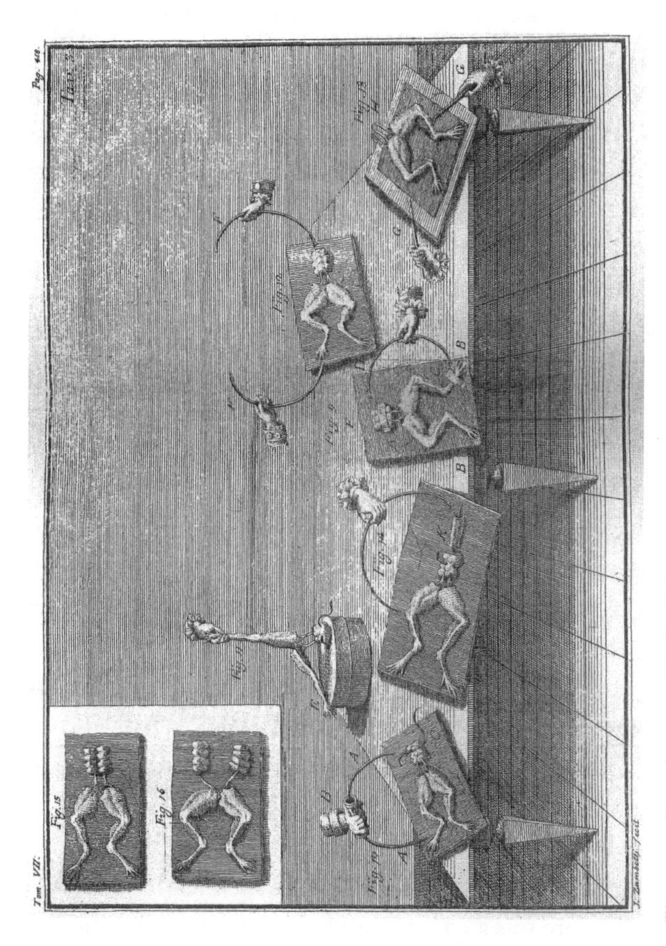

Figure 2. Experiments with frog's legs.

Luigi Galvani was born in Bologna on 9 September 1737. He graduated in medicine from the University of Bologna, where he studied anatomy under Domenico Galeazzi as well as physiology. In 1762 he took up a post teaching anatomy in the University, a job that was coupled with a professorship in obstetrics in the Institute of Arts and Sciences. His studies of animal electricity emerged as just one among his many-sided medical researches.

He married Lucia Galeazzi, the daughter of his anatomy teacher, in 1764. Just as Madame Lavoisier had served as a laboratory assistant in her husband's researches, so too did Lucia for hers. There is a legend that connects her intimately with the origins of the experimental programme around frog's legs. The story has it that she had fallen ill, being of a 'delicate constitution'. The uxorious Luigi was preparing frog's legs as a delicacy for his ailing wife, when another assistant perhaps playfully touched the nerve endings of the yet uncooked material with a scalpel. This produced the spasm, the physiological effect that initiated the research. Charming though this tale is, it cannot be true. Unless the kitchen was doubling as a laboratory the coincidence that a second assistant should have noticed that the effect was coincident with the production of a spark on a friction generator some distance from the legs was too unlikely. Much more likely is the alternative tale that Lucia was herself the first assistant, the one who touched the nerves with a scalpel.

Far from concealing the incomplete and tentative nature of the work, Galvani begins his famous book with a modest disclaimer. Though fortune played a role in the work, the point of publication was to 'light the way for those who have wished to embark on the same course of experiment'.[4] Using other 'prepared' frogs, he soon eliminated the possibility that it was the point of the scalpel that was the effective stimulus. 'The phenomenon occurred, however, only as often as a spark was produced.' The proof of the electrical nature of the phenomenon was strengthened by the observation that holding only the bone handle of the scalpel effectively insulated it, and no effect occurred. When the fingers touched the blade or the rivets securing it the effect reappeared.

In a series of trials with other means for applying electrical stimulation to the crural nerves not only of frogs but of chickens and sheep, it became clear that the active ingredient was electricity, both 'natural' drawn from thundery weather, and 'artificial', generated in the laboratory by a friction machine. At this stage, the dissected limbs were the focal point of the research. Under what conditions and with what external connections did the effect occur? The problem was to decide whether animal spirits were of the same nature or merely of a similar nature to the various forms of electricity then known. However, as the programme progressed, the focus shifted to a study of the means by which electricity could be generated. In the new context, the frog's legs took on the role of instruments, detectors of electrical force.

The original observation that set Galvani off on his experimental programme was the coincidence between three events, touching the exposed crural nerve with a grounded metal object, the muscular spasm, and the spark produced by a friction generator some distance away. Analogies or models almost always dominate scientific thinking. In this case Galvani's belief in an autonomous 'animal electricity', similar to common electricity, led him to an erroneous model for the electrical nature of muscles. He began to think of a muscle or group of muscles as a kind of animal Leyden jar. The Leyden jar was the first effective electrical condenser, a device for storing 'electric fluid'. It consisted of a glass or earthenware jar with a metal rod passing through the lid. By connecting the rod to a generator electricity could be 'condensed' and stored in the jar. Each muscle or muscle group, an 'animal Leyden jar', so he thought, was originally charged with animal electricity, which was discharged by the touch of a conductor such as the metal of the scalpel that had activated the spasm in the original observation. The frog's legs were organic Leyden jars, inside and outside differing in polarity (74).

The Leyden jar model was not entirely mistaken. The whole animal sometimes played the role of condenser. In 1779, Lord Mahon had described the phenomenon, calling it the 'returning stroke'. Farmers had noticed that an animal could be electrocuted during a storm, even if the lightning had struck some distance from it. The animal became electrically charged by induction, a process not clearly identified in the eighteenth century. The discharge of the electricity stored in the animal as a condenser when the creature was earthed killed it.

However, in the case of the returning stroke the fatal electricity was not originally present in the animal, but induced by the lightning strike. Galvani's mistake was to suppose that animal electricity was naturally present in the muscles as in charged Leyden jars. The touch of the conducting scalpel or wire, he thought, discharged the stored animal electricity by completing a circuit between positive and negative polarities.

Galvani's misunderstanding was quite systematic. He begins by asserting that 'the muscle is the appropriate seat of the electricity we investigated, with the nerve functioning as a conductor.' He goes on to say that 'just as a discharge is produced in a Leyden jar when an [metallic] arc is applied to its conductor . . . electric fluid can flow and produce contractions through a nerve constructed in a similar way [to a conductor].' Since he thought that the muscular contraction was produced by electricity already stored in the muscle he remarks that 'animal electricity . . . has need of no preliminary device like rubbing... to excite it, but being prepared, as it were, to manifest itself immediately it reveals itself merely by contact.' All we need is 'something that summons the nerve-electric fluid from the muscle to the nerve and provokes its flow'.

Another of Galvani's discoveries transformed the basis of electrical science itself. We take for granted that electric current can be generated mechanically, as in the generators that power the national grid, and chemically, as in the batteries that drive a huge range of domestic devices. Cars make use of both methods of creating and storing electricity, having both batteries and generators. Galvani made the fundamental observation that led eventually to the invention of the electric battery in which chemical reactions are the source of electricity.

The hypothesis that lightning was a form of electricity had been established around the year 1752 by a series of experiments initiated by Benjamin Franklin and carried forwards in France. By erecting a long bar of iron pointed at the tip, Thomas-Francois D'Alibard was able to observe the 'same kind of commotions [sparks of fire] as in the common electrical experiments' when a storm cloud passed overhead. Michel De Lor went further with a bar ninety-nine feet in height achieving even more satisfactory effects, 'the same noise, the same fire, and the same crackling'.[5] Galvani set up a similar apparatus, a pointed iron bar mounted on a high point of a building and connected to a wire, to try the effect of 'atmospheric electricity' on his frog's legs. The same spasms as had been set off by metallic earthed contact were also brought about in this manner by atmospheric electricity.

Carrying forwards the study of the relation between muscle stimulus and lightning, Galvani hung a pair of prepared frog's legs on an iron bar by means of a copper hook. The legs twitched in the expected manner when lightning struck nearby or even when storm clouds were about. However, to his amazement, the legs twitched when there was no sign of the presence of atmospheric electricity in any form. They say that accidents are fruitful food for the prepared mind. Galvani realized that it was the rubbing of the copper hook on the damp iron grill that was implicated in the stimulus of the legs. Since that stimulus was electrical, the rubbing of the dissimilar metals must be producing electricity, in sufficient quantity to bring about the discharge of the pre-existing animal electricity in the organic Leyden jar that he conceived the pair of frog legs to be.

Following his erroneous theory Galvani set about experiments with the frog's legs detector, using arcs of various metals and combinations of metals to earth the legs. He found that non-conducting arcs produced no effect, nor did arcs made of just one metal. But with arcs with sections of different metals the effect reappeared. He failed to see that the bimetallic arc was the source of the electrical stimulation, not a path for the discharge of 'nerveo-electric fluid' from the 'Leyden jar' muscle.

He concluded that the metals must be different, and that the contact must be damp. And there Galvani's researches into the chemical production of electricity ended. Throughout this programme frog's legs served as the instrument for detecting electricity produced by the contact of dissimilar metals. Galvani made no attempt to quantify the effect, except to note some crude differences in the violence of the muscular spasms stimulated by contact with different sources of electricity and different metallic combinations.

This great discovery was exploited by Galvani's flamboyant and energetic rival and critic, Alessandro Volta. In order to detect the weak currents produced by the contact of damp metals he needed a very sensitive electroscope. Appropriating not only Galvani's discovery but his instrumentarium as well, he adopted the frog's legs as a detector. The sensitivity of this organic electroscope was quantified by Volta as about 40 times that of the metallic instruments he then had at his disposal.

Galvani's life, already devastated by the death of Lucia in 1790, was fatally disrupted by the political events of the time. The French had driven out the Austrians who had hitherto ruled northern Italy. In 1797 Napoleon set about creating a Cisapline Republic, with himself as the first President. Bologna fell within the boundaries of the new French-sponsored state. Galvani, as the Rector of the University, was required to swear an oath of allegiance to the new regime. This he refused to do. Accounts of his life do not make it entirely clear why he took such a damaging stand. Since he was conservative by temperament the necessary pragmatic accommodation to the reality of the political situation was beyond him. He was dismissed from the University, and the Istituto. His life dragged on for a few months in poverty and neglect. He died in 1798.

Galvani's discovery of 'chemical electricity' depended on having an electroscopic device sufficiently sensitive to detect the tiny currents produced by the rubbing of a copper hook on an iron bar. The frog's legs provided the right degree of sensitivity. So we see a shift of focus. In the initial studies the behaviour of the muscles was the research object. When the serendipitous observation of the chemical effect was followed up, the legs became the instruments by which electrical currents could be detected, taking the neuroelectrical effect for granted. Since the effect alone was what was required to use the legs as a scientific instrument, the fact that Galvani held to a wholly erroneous theory as to the nature of the effect did not matter, at least at the time.

Indicators of acidity and alkalinity

Perhaps the one piece of information most of us carry away from science courses in school is the fact that litmus paper turns red when dipped in an acid solution and blue in one which is alkaline. Behind this commonplace phenomenon lies an interesting history. The active ingredient of litmus comes from a kind of organism, the lichen, the greenish-greyish stuff one finds on the surfaces of rocks. It is also the basis of the technique of lichenometry, using lichens to determine the age of rocks, to be described in Chapter 3.

When a substance dissolves in water it usually splits up into a positively and a negatively charged part, a positive and a negative ion. Simplifying the chemistry somewhat we can say that acids, such as hydrochloric acid, HCl, always yield positively charged hydrogen ions (H^+) when dissolved in water. Alkalis, such as sodium hydroxide, NaOH, always yield negatively charged hydroxyl ions (OH^-). The standard way of representing degree of acidity and alkalinity is by the pH number. A neutral solution has a pH 7. The degree of alkalinity is denoted by pH values greater than 7, while the degree of acidity is denoted by pH values less than 7. When the molecules of litmus interact with H^+ ions their electron configuration takes a different form from that induced by interaction with hydroxyl ions (OH^-). The diffraction of light through molecular structures depends on the electronic configurations. The former configuration yields light of longer wavelengths (looking red) while that of the other electron configuration yields light of shorter wavelengths (looking blue).

Chemical indicators display one colour in alkali solutions—litmus turns blue—and another in acidic solutions—litmus turns red. Many other naturally occurring substances change in a similar way. For example, hydrangea and rose petals display different colours when in alkaline and in acid solutions. The active ingredients of many such 'natural indicators' have been isolated. But what of litmus?

Lichens are curious beings. They are compound organisms, consisting of a symbiotic union between certain species of fungi and unicellular green algae. Slicing through the thallus or disc of a lichen discloses a white fungal layer enclosing a thin green layer of algae.[6] The chlorophyll in the algae produces carbohydrates by photosynthesis which sustain the fungus. The fungus provides a secure home place for the algae. This combination allows them to thrive in otherwise very hostile environments, such as the surfaces of exposed rocks.

Lichens have long been a source of dyestuffs. For example, orcein has been used for millennia as a purple dye. More recently it has been used as a stain for revealing the elusive internal structures of bacteria. Its relevance to litmus comes from the fact that the substances in the lichen from which it is derived are also the precursors of litmus. The extraction of litmus from lichens involves a separation from red orcein dyes.

Our knowledge of these two related substances is due to the life-long studies of just one man, Hans Musso (1925–88). He was Professor of the Organic Chemistry Institute at Götingen University from 1969 until his death. During the 1950s and 1960s he traced out the nature of many of the constituents to be found in orceine and its relative litmus. These were based on a structure to be found very widely in organic chemistry, the hexagonal pattern of carbon atoms, the benzene ring.

The practical recipes used by the manufacturers of these dyes and indicators depended on ammonia, a pungent product of the breakdown of urine. The extraction of litmus from another lichen, *Roccella*, follows a similar line, but with the addition of lime to the process.

A human story lies behind the ready availability of litmus paper today. The active ingredient is extracted from the lichen *parelle*, originally a Dutch monopoly, by using a secret recipe. However, in 1841 the French chemists C. L. and A. B. Berthollet published a method for extracting dyes from lichens, including *Roccella*, a common source of litmus. The best lichen for the purpose of litmus extraction is *Rocella montagnei* found commercially in Madagascar. The method was complex but based, as I remarked earlier, on the use of urine and lime. A modern version of the recipe runs as follows:

The lichens are first soaked in a solution of sodium carbonate. Ammonia is then added. Stir the lichens from time to time for four weeks. 'The colour changes from red to purple and finally blue. The lichens are then dried and powdered. At this stage the lichens contain partly litmus and partly orcein pigments. The orcein is removed by extraction with alcohol, leaving the pale blue litmus.'[7]

The Dutch monopoly of the litmus trade was finally broken in 1940 by Johnson's, a British company. The date must surely signify that the German invasion of Holland in 1940 had stimulated this move. Just as the threat of being cut off from foreign petrol supplies stimulated German research into the synthesis of petrol from coal, so the drying up of Dutch litmus supplies no doubt encouraged Johnson's to go into the business of its manufacture.

Local weather forecasting

Anthropologists have persuaded us that many of the practices of people living outside the European sphere or before the time that controlled experimentation became widespread ought not to be dismissed as magic or superstition. They are often expressions of practical rules and have some claim to be taken seriously as folk science. Some, however, such as the use of the entrails of birds to make predictions about the future have not stood the test of time. Examples of the uses of living creatures and their parts for detecting the state of the environment fall into two groups. There are those that have been used just to monitor the state of things in the here and now. Of growing interest are another group of devices which can be used to detect changes thought to have ominous consequences for the future.

A report in the *Guardian* newspaper for 4 December 2004 by Owen Bowcott mentioned two folk predictors of a hard winter to come. Flocks of Siberian Waxwings had been seen in Scotland and the eastern areas of England, indicating that cold weather was to be expected. The report also mentioned the traditional claim that when the holly bears an exceptionally large crop of berries in the autumn one should expect a cold winter to come. While these long-term forecasts should be treated with some scepticism, the behavior of insects can be a good short-term predictor of weather.

Clouds of greenfly, black fly, and aphids, very small insects, are drawn up sometimes as far as a kilometre into the atmosphere locally in just those weather conditions that produce sudden showers. These insect clouds show up on local radar before the radar displays the signs of the falling of an actual shower. The images of insect swarms can be distinguished from random radar echoes and so can be detected prior to sudden downpours. The insects are drawn up into the atmosphere by updraughts caused by cooler inshore breezes meeting warmer inland winds. According to Anthony Illingworth of Reading University 'the leading edge of clear, cooler air from sea breezes sweeps up insects like a brush.' Here we have a living detector of an imminent shower.

Detecting pollutants in the environment

Current interest in the environment has led to the development of a great many devices for monitoring the current state of the atmosphere and the Earth's watery resources, seas, rivers, and lakes. Some of them are very simple, such as the use of certain plants sensitive to pollutants. Others are very sophisticated such as the transgenic fish that glow in the presence of contaminants in the water.

MILKWEED AS A DETECTOR OF OZONE

Excess levels of ozone in the lower atmosphere are generally regarded as harmful. The presence of O_3 can be detected by examining the leaves of the milkweed plant. Several species of this weed are found extensively in North America. The plant is very sensitive to this common pollutant. If there is an excess of ozone the leaves are stippled with small dark sharply edged patches. The more dense the spots, the higher the ozone concentration.

CANARIES IN THE MINE

The use of canaries to detect dangerous gases in coal mines is not exactly a contribution to science as we ordinarily think of it. However, this is such a fascinating story that it deserves a place nevertheless. Mines are dangerous places. They are made more hazardous by the presence of gases such as methane or 'fire damp' and carbon monoxide (CO), which was not only inflammable but poisonous as well. It is odourless, tasteless, and colourless. In the days when the only illumination in a mine was candlelight, the possibility of setting off a fatal explosion was very real. Humphrey Davy's invention of the miner's safety lamp was one safeguard against igniting these gases, but it did not provide any way of detecting them. Neither carbon monoxide nor methane can be detected by the human senses, so a more sensitive instrument was needed. Mice were commonly used to warn of the buildup of carbon monoxide since they showed signs of distress in those circumstances. However, it was found that canary birds were more sensitive to this danger, and displayed more visible symptoms. According to the US Mine Safety Health Administration canaries stopped singing and swayed on their perches when even a small amount of CO was present.

Canaries were introduced into British coal mines in 1911, replacing the mice that had previously alerted miners to the presence of dangerous gases. Two canaries were provided for each pit. Cramped in their small cages they went down with the miners. These birds were much cherished, and astonishingly they continued to sing under the unfavourable conditions below ground. However, in 1986 the canaries were phased out in favour of electronic detectors.[8] It is said that these living instruments were much missed by the miners, despite the sophistication of their new non-living detectors.

TRANSGENIC FISH TO DETECT WATER POLLUTANTS

Plants and plant products are not the only organic beings that respond to specific conditions in their environments. Transgenic versions of Zebra fish have been bred to display different fluorescent colours when they encounter specific toxic substances in water supplies.

The Zebra fish (*Brachydanio rerio*) is a native of India. It was first collected and brought back to Europe as a decorative addition to the aquaria so popular in the mid-nineteenth century. In its native form this tiny fish is strikingly coloured in longitudinal stripes of black and silver. In its natural habitat the zebra fish does not display any remarkable characteristics, such as colour changes in response to changes in the environment. However, for the past decade or so this creature has been the focus of intense research into the expression of genes in the development of the embryo. As a convenient experimental subject it has been used in laboratories all over the world, including Harvard and Singapore. For example in a massive programme of studies of zebra fish at University College, London, there are ten research teams at work covering topics from the development of patterns in the forebrain to the embryology of the circadian clock—the regulator of daily cycles in the body of the fish. There is even a journal devoted exclusively to studies of this creature, *Zebrafish*.

Techniques for gene insertion from one organism to another are now highly developed. Among various applications in the field of zebra fish research has been the creation of transgenic lines with distinctive displays of colouration as an associated response to various chemicals in the environment. Merely having a particular gene in the genome is not sufficient to cause a propensity to change colour to appear in the adult organism. A genetic 'switch' must be turned on so that the gene will express itself in sensitive proteins in the body of the adult fish. The colours appear in response to induced changes in the chemical make-up of the skeletal muscles.

A transgenic organism is one which develops characteristics that come from a gene or genes inserted into its DNA, genes taken from that of another creature which has the relevant characteristics as part of its natural endowment. The process involves *transferring* a gene or genes from one organism to another.

Genes are specific sequences of the base pairs that make up the double strand of DNA. They control the production of specific proteins. A stretch of DNA that encodes the molecular plan for the protein is copied by assembling many matching sequences of base pairs, the RNA, which spread through the cell and control the synthesis of the specified protein. However, cells must produce proteins at the right time in the right quantity for the job in hand. Each gene sequence is preceded by a sequence of base pairs that serve as *promoters* for the initiation of the process of RNA replication and protein production elsewhere in the cell, the process of *gene expression*. The promoter is activated by a stimulus from a molecule that comes from somewhere in the organism, perhaps originally from the external environment, bearing tidings of some present need, so to speak.

The transfer of genes from the cells of one organism to those of another requires two procedures. First of all the gene to be transferred must be isolated from the DNA molecule in which it occurs naturally. This involves separating a sequence of base pairs from the original DNA. Next the isolated fragment of DNA must be inserted into the DNA of the target cell in such a way that it then replicates and the new gene spreads throughout the target organism.

The cutting out of a DNA sequence is accomplished by a *restriction enzyme*. These enzymes have evolved to protect cells from the incursion of foreign DNA. Such enzymes 'recognize' sequences of base pairs in the DNA of various lengths, 4, 6, or 8 base pairs. The enzyme actually breaks the DNA strand into single strands which can be joined to other fragments of DNA.

Fragments which can rejoin others are said to have 'sticky ends'. Sometimes the restriction enzyme produces double-stranded sections. Both kinds of fragments are joined together by another enzyme, a *ligase*.

The second procedure involves the use of a *vector* to carry the recombined DNA to its new site in the target cell, and plasmids are used for this. Plasmids were discovered in the course of research into the mechanism by which some bacteria are resistant to antibiotics. Plasmids are small loops of DNA molecules found outside the chromosomes of the nucleus. Plasmids are made up of the same four bases, but they might contain thousands of base pairs compared to the millions in chromosomal DNA. By inserting plasmids containing the alien gene and its promoter sequence into the cells of the target organism, the familiar process of protein production by RNA copying begins with the DNA of the plasmid as the template. By inserting genes that code for proteins sensitive to the substance to be detected into zebra fish, living transgenic detectors of pollutants can be created. The protein molecules sensitive to the pollutants are constituents of the muscle fibres of the adult fish. When the pollutant is taken up the muscle fibres fluoresce.[9]

Dr Zhiyang Gong has led a team at the University of Singapore that has successfully produced fish which respond to various pollutants such as the hormone oestrogen as well as to poisonous metallic contaminants. This is part of the field of 'toxicogenomics'. According to Dr Gong his research team has produced 'dozens of transgenic fish lines' for all sorts of purposes.[10] Most of the work at the University of Singapore has been devoted to developing the fish as models for tracking the development of cancers, for example cancer of the liver.[11] The indicator version of the fish is another transgenic line that has been created by the use of standard genetic engineering techniques. Zebra fish eggs can be injected with the colour-determining gene of a red sea anemone. The mature transgenic fish turn from their original black and silver to fluorescent red in the presence of specific pollutants. By using a gene from a fluorescent jellyfish a green response is established in the adult fish, making it available as a detector of a number of different contaminants of water supplies.[12]

If these genetically engineered fish were to be released into the environment to act as devices to monitor the state of the water in some area would they tend to replace the wild species? According to Dr Gong, the metabolic demands made on the organism by the fluorescent colour display, and the greater visibility of such individuals to predators would mean that these characteristics would not be advantageous biologically. The added genes would not spread through the native population by natural selection.

Not surprisingly, the availability of new kinds of brightly coloured fish has opened up commercial opportunities. Who would not want a glowing red or green specimen in their home aquarium? One of Dr Gong's recent papers significantly includes the word 'ornamental' in its title.[13] The commercial exploitation of this technique has already begun to flourish.[14]

Detecting climate change

Plants and animals used for detecting and measuring temperature, humidity, rainfall, and prevalence and strength of winds are the most important instruments for the study of climate change. Not only are they the most important for tracking what is happening now, but it is clear that they are among the most powerful devices by which the study of the details of ancient climates is made possible. Before systematic correlations of temperature measurements around the world began, global warming had manifested itself in changes in the distribution of certain plants and animals. By recovering evidence of these changes we can get some idea of how the climate of the Earth has changed over time.

Records of the physical parameters of climate began in the seventeenth century in Western Europe. A few documents survive from three or four thousand years ago describing local phenomena that can be used as evidence for changes in climate. Palaeontology will take us only so far—certainly the fossil record allows geologists to infer climate changes deep into the Earth's past. Though much can be learned that way, a certain amount of fine detail comes from palynology, the science of the use of pollen as an indicator of the state of the climate at certain well-defined times. Palynology has been refined to become much more than a way of detecting climate change; it is also a device for measuring climatic eras and dating the moments at which these changes have occurred (see the discussion in Chapter 3).

NOAH'S ARK

The first recorded example of the use of plants and animals as means to detect climate change must surely have been by Noah. There is a graphic description in the first book of the Bible.

> And the flood was forty days upon the earth, and the waters increased, and bare up the ark, and it was lift up above the earth. . . . And the waters prevailed exceedingly upon the earth: and all the high hills; and the mountains were covered. Fifteen cubits upward did the waters prevail; and the mountains were covered.

Noah's use of plant materials to detect the presence of dry land is described as follows:

> And he sent forth a raven, which went forth to and fro, until the waters were dried up from the earth. And he sent forth a dove from him, to see if the waters were abated from the face of the ground. But the dove found no rest for the sole of her foot, and she returned unto him in the ark, for the waters were on the face of the whole earth; then he put forth his hand, and took her and pulled her unto him into the ark. And he stayed yet other seven days; and again he sent forth the dove out of the ark. And the dove came in to him in the evening; and, lo, in her mouth was an olive leaf plucked off; so Noah knew that the waters were abated from the earth.[15]

Here is the principle of such studies as palaeobotany vividly illustrated. If there are olive trees from which a leaf must have come, then there must be dry land and the floods have drained away. The evidence of climate change was overwhelming.

WILLIAM BUCKLAND AND THE HYENA'S CAVE

To understand the significance of William Buckland's account of the bones in the famous 'Hyaena's Den' found in a cave in Kirkland in the Northern England county of Yorkshire we need to bear in mind the widespread belief in the eighteenth and nineteenth centuries among scientists and biblical scholars alike that there had been at least one major inundation. The Flood was widely assumed to have been a real historical event. The Earth, or at least major areas of its surface, had been covered by water. The story of Noah sufficed for the pious, while the distribution of gravel beds, the scattering of stones far from their places of origin, and the evident of forceful shaping of valleys supported geologists in the 'diluvial hypothesis'. Recent geological eras were divided into the antediluvian and postdiluvian, before and after the Flood. Some thought there might have been several floods of which we see the consequences only of the most recent. The distribution of gravel and rocks across the surface of the earth had been the effect of a great 'diluvial tide'. This popular theory was to be challenged by Charles Lyell (1797–1875) and others. Observations of valley configurations and the layering of volcanic ash above that of 'diluvial' gravel gradually undermined diluvialism.

But this theory prompted a question worth considering concerning the climate of particular places on the surface of the Earth before and after the Flood. Had the climate changed? Had the antediluvian climate of England been similar to the climate in the nineteenth century? Buckland's careful interpretation of the bones found in the Hyaena's Den and their current state provided a convincing answer to the question. The climate, Buckland declared, had changed. The hyaenas had been extinguished by the Flood. And after the Flood the climate did not encourage their return.

William Buckland was born at Axminster in Devon on 12 March 1784.[16] His father, Charles Buckland, was the rector of two country parishes. William was educated at Tiverton School near his home, and later at St Mary's College in Winchester. As a child he was fascinated by the fossils revealed by excavations in the local quarries. A scholarship took him to Corpus Christi College at Oxford. He was elected to a Fellowship at Corpus in 1808, and, as was the custom in those times in Oxford, ordained as a priest in the same year. He had already discovered an abiding interest in geology. The Professor of Mineralogy, Dr Kidd, had popularized geology in his lectures and had succeeded in interesting the University in the new science. Buckland had evidently made a very favourable impression, so much so that in 1813 he became Kidd's successor as Professor of Mineralogy. In the years after 1808, influenced by the work of William Smith made known to him through several common friends, he rode over much of south-west England collecting samples of strata and their characteristic fossils. He extended these wanderings not only into the rest of England but into Wales, Ireland, and Scotland as well. Later expeditions took him to Germany and Switzerland.

A popular and lively lecturer, he included not only mineralogy but geology in his annual presentations. This was one amongst many influences that helped to promote great public interest in earth sciences at the beginning of the nineteenth century. Only five years after his appointment to the Chair of Mineralogy he was made Reader in Geology, a post created especially for him. In the same year he was elected a Fellow of the Royal Society. His investigation of the Hyaena's Den in 1821, published as *Reliquiae Diluvianae* in 1824, was one of the most potent influences on the growth of geology and palaeontology throughout Europe.

He is probably to be credited with the first discovery of a dinosaur in an excavation at Stonesfield Quarry near Oxford in 1824. He was appointed to a country parish in 1828, a necessary move if one intended to marry. College fellows were supposed to devote their whole energies to the job. However, in the same year he was appointed a Canon of Christchurch, a plum job in Oxford, and shortly thereafter married Mary Morland. He never became a parish priest. Mary was a competent geologist and after her marriage continued to illustrate scientific books, particularly geological publications.

In 1838 on a visit to Switzerland Buckland noticed the effects that glacier movements had on the rocks and indeed their effect in shaping the landscape. Returning to England he began to apply this idea to the English landscape, in conscious opposition to the diluvial theory of his youth.

His work received well-deserved public recognition. He became Chairman of the Geological Society in 1824, having been awarded the Copley Medal of the Royal Society for his work on the Hyaena's Den. To his surprise he also advanced in the Church, becoming Dean of Westminster in 1845. Troubled by illness, perhaps the result of his extraordinary practice of eating all sorts of wild creatures, and insisting that his guests did too, he retired to Islip near Oxford, where he died on 4 August 1856.

What was it that this industrious and amiable man found in the Hyaena's Den? In the course of excavating for a quarry some workmen had broken into a cave, the floor of which was covered in bones. They simply scattered these over the local road. A passer-by realized that these were relics of some geological significance. Very soon word had reached William Buckland in Oxford. He visited the site immediately and made a very thorough examination of the cave itself and its remaining contents.

The cave had been effectively sealed by local earth movements and vegetation. This suggested to Buckland that the bones in the cave could not have been swept in by the rush of waters of the diluvial tide. Among the bones were those of hyaena,[17] elephant, rhinoceros, hippopotamus, horse, ox, deer, bear, rats, and birds. In a letter of 1821 Buckland declared 'we can only suppose the bones to be the wreck of animals that were dragged in for food by the hyaenas.'[18] The bodies of elephants, rhinos, and hippos were far too large for these animals to have wandered in alive and simply died in the sanctuary of the cave. To support the hyaena theory he pointed out that 'on some of the bones, marks may be traced, of the canine teeth of the hyaenas that occur in the cave.' He cross-checked this observation with experts on the markings left on bones by various carnivores.

'It must already appear probable', he wrote, 'from the facts above described, particularly from the comminuted state and apparently gnawed condition of the bones, that the cave at Kirkdale, during a long succession of years, was inhabited as a den by hyaenas, and that they dragged into its recesses the other animals' bodies whose remains are found mixed indiscriminately with their own.' This was further supported by the presence of balls of excrement characteristic of hyaenas, and which were especially rich in matter derived from bones. It should also be added that the bones had not become petrified, but simply overlaid with calcareous deposits from the stalagmites that formed in the cave. They were preserved in an almost perfect state.

Now for the step from the bone fragments to the claim that the climate of Northern England had changed. Since there was no sign of diluvial gravel in the cave, the bones had been protected from the flood. They were antediluvian: 'the facts developed in this charnel house of the antediluvian forests of Yorkshire demonstrate that there was a long succession of years in which elephant, rhinoceros and hippopotamus had been the prey of hyaenas, which like themselves, inhabited England in the period preceding the formation of the diluvial gravel.' These animals are now confined to the warmer southern latitudes. Since it seems that those whose remains were preserved at Kirkland lived and died in the regions where their remains are now found, 'that the animals whose remains are found interred in the wreck of that inundation [the final flood] were inhabitants of the high north latitudes', the climate of Yorkshire must then have been tropical. This was further borne out by 'the abundant occurrence of fossil crocodiles . . . palms . . . nearly allied in structure and character to those which are peculiar to hot climates'. The climate of Yorkshire had changed substantially. The phenomenon had been detected, but it could not yet be measured in any reasonably accurate way. No one knew when the change in climate had occurred though there were plenty of educated guesses. Popular opinion put the Flood about four thousand years ago. Yorkshire's climate must have been tropical at least that long ago.

The cave and its contents are a device for detecting climate changes that have taken place over fairly long stretches of time. Buckland himself estimated that the bones had lain there for at least four thousand years, that is, from the time at which it was presumed that the Flood had sealed off the cave.

Buckland's studies put paid to another popular theory, that the Flood, the fatal tide, had been the result of catastrophic collapse of existing dry land and the raising up of the seabed. This could not have happened, Buckland reasoned, because Yorkshire had been dry land before and after the flood tide that had killed off the tropical species. He refused to speculate on the origins of the great inundation. He was not ready to declare with certainty that it had never happened.

Detecting dangers in the Universe at large

What is the nature of the regions of space above the surface of the Earth? Astronomers, meteorologists, and others have worked out all sorts of answers, relying on their Earth-bound observations. Were the conditions 'up there' inimical to human life? One doesn't have to think too adventurously to be worried by the possibility that the Universe at large is hostile to life. Two recent major space expeditions were designed to discover the nature of distant objects. The atmosphere of Jupiter was assessed by dropping a space module with suitable instruments which sank down through the outer envelope of gases until it was destroyed. Another experiment involved crashing a probe into a comet to use the debris of the explosion to investigate its composition. Finding what something is made of is a special case of scientific detection. In both these cases the 'instruments' were not recoverable. Other expeditions have involved scooping up samples from distance astronomical objects and bringing them back to earth. However, they were all inorganic probes, made predominantly of metal. The stories to be told in this section involve the return, recovery, and sometimes loss of instruments specifically chosen to detect influences hostile to life. What better than living beings?

THE UPPER ATMOSPHERE

The first time conjectures about the nature of extraterrestrial space were tested experimentally a small menagerie of living creatures was sent aloft. A duck, a rooster, and a sheep ascended in Paris while two hundred years later dogs were favoured by the Russian space programme.

It would not be an exaggeration to say that the successful launching of a hot air balloon under controlled conditions was the beginning of aviation.[19] Joseph and Etienne Montgolfier were the twelfth and fifteenth children respectively of the wealthy French paper manufacturer Pierre Montgolfier and his wife, Anne. They differed greatly in temperament. Joseph, the elder by five years, was inventive, mechanically dexterous, and curious. He preferred self-education to schooling. As a youth he set up a small dye-stuffs business. Etienne, the younger, was attentive to schooling and obediently joined the family business. Like other couples who have made great contributions to science and engineering, such as Michelson and Morley, the researchers who showed that speed of light was always constant, one was something of a maverick and the other staid and steady.

Joseph developed the idea of flight by a hot air balloon in the course of a programme of studies on the general relation between heat and mechanical work, during the years 1777 to 1783. At the time, heat was thought to be the effect of a substance, caloric, that was emitted in combustion. Lavoisier, for example, listed it among the chemical elements.

To test the 'levity' of caloric Joseph Montgolfier filled a small box-like envelope of wood and cloth with hot air from a fire. The box was a mere 3 ft square and 4 ft in height. It rose swiftly to bump against the ceiling. Joseph knew how to calculate the levity of such a balloon by calculating the difference in weight of original air and of the 'caloric-imbued' air that replaced it as the balloon became filled with the hot gases from the fire.

Recruiting Etienne to the project, he set about constructing larger and larger balloons of cloth lined with paper. Abandoning the idea of a framework, the brothers realized that the hot air would expand the balloon to whatever shape and size was, so to say, built into the envelope. They made successful trials with 'flexible bubbles' in April 1783. There were problems with rain tending to damage the envelope, but, nevertheless, in general, up into the sky went the balloon, sometimes travelling for several kilometres until the cooling of the air inside the balloon brought it down.

Anxious to beat their rivals, who were constructing hydrogen-filled balloons, the Montgolfiers organized a grand demonstration at Versailles for the King (Louis XVI destined for the guillotine) and his equally ill-fated Queen, Marie-Antoinette. This balloon was 70 ft in height and 40 ft in diameter, richly decorated. Joseph knew that it could easily carry passengers as he had seen the men holding the guy ropes lifted into the air as the balloon filled. The demonstration was a huge success.

The next step would be to attach a gondola with a small stove to produce more caloric for a sustained flight with an aviator on board to control it (Figure 3). But who knew what the air was like at a thousand metres above Paris? Perhaps at that height a human being could not survive. It is said, with what authority I do not know, that the King himself forbade manned flight until the nature of the higher atmosphere should have been tested by sending up a balloon with a selection of living creatures and bringing them safely back to earth. In any event that is what Joseph and Etienne did.

There was evidently some discussion as to the choice of animal to serve as an instrument to detect any malign influences in the higher reaches of the atmosphere. According to Charles Gillispie, Etienne Montgolfier's first choice was for a sheep.[20] It was suggested that a dog would be preferable since it would make itself heard should anything untoward occur. In terms of instruments it would give a much more legible readout. Alternatively why not a cow or even a young bull? Joseph, with an eye on publicity, argued for a cow. He insisted that this would have an extraordinary effect, far more so than a panicky sheep which no one would be able to see. They finally settled on a sheep, a duck, and a rooster as the first aeronauts.

The flight was evidently a great success not only technically but in terms of publicity. A huge crowd assembled at Versailles to see the show. The animals returned unharmed. Etienne reported their condition in a letter to his wife in the form of a spoof newspaper report of the comments of the aeronauts.

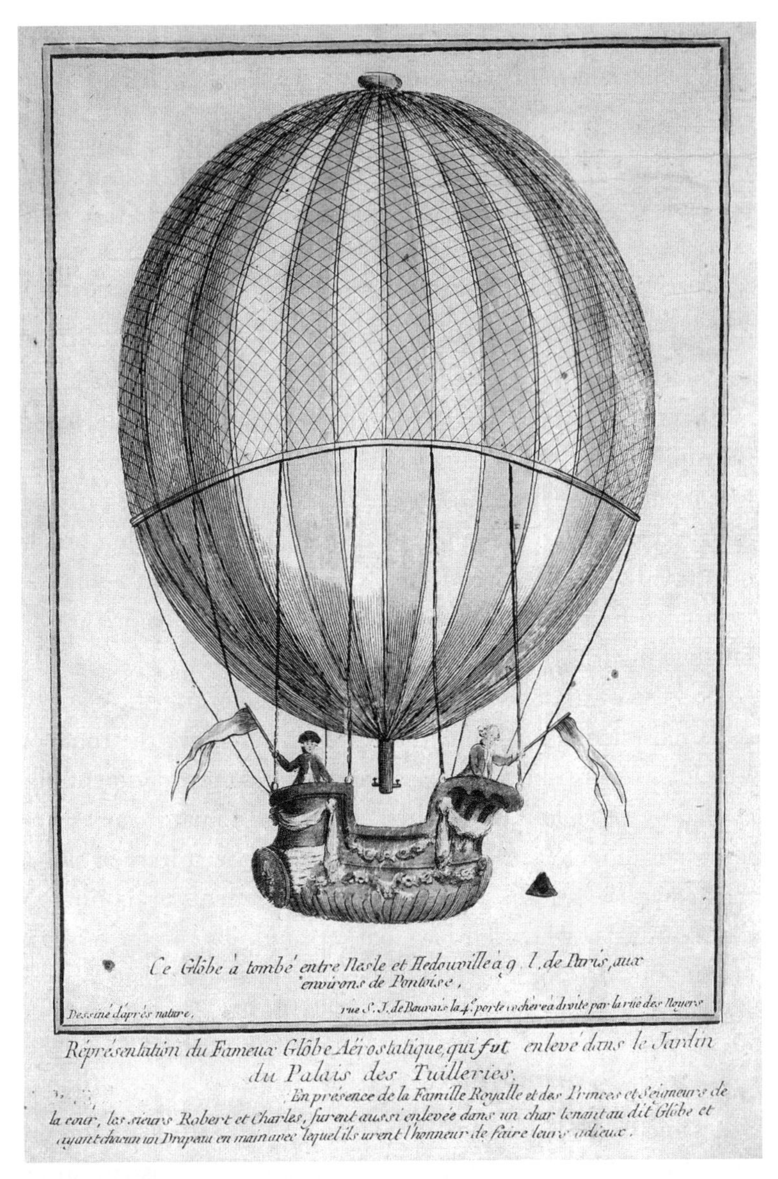

Figure 3. The Montgolfiers' balloon.

News from the Air, 19 September 1783. ***Aboard the Aerostat Réveillon***

> 'We feel fine. We've landed safely despite the wind. It's given us an appetite.' That is all we could gather from the talk of the three travelers, seeing that they do not know how to write and that we neglected to teach them French. The first could only say 'Quack, Quack'; the second 'Cock-a-doodle-doo'; and third, no doubt a member of the Lamb family, replied only 'Baa' to all our questions.[21]

With safety assured by the favourable response of the living instrumentation to the 'higher' reaches of the atmosphere, the first manned free flight was made by Pilatre de Rosier, carrying man and stove into the atmosphere, high above the tallest buildings in Paris. De Rosier had already spent some time in a gondola suspended from a tethered balloon. He was accompanied in the first free flight by the Marquis d'Arlande.

The use of animals as instruments for detecting the state of the atmosphere a mere one thousand metres above the roofs of Versailles may strike us not only as a redundant experiment but as slightly ridiculous. Yet at that time this domain was as unknown as outer space. Of course, much had been learned about the higher levels of the atmosphere by experiments in mountainous regions. Yet who was to say that the conditions on the summit of the Puy-de-Dome accurately represented the state of the atmosphere at that altitude above the plains?

THE CONDITIONS IN OUTER SPACE

The same reasoning as prompted Louis XVI to urge caution on the Montgolfiers animated the planners of the Russian space programme. Russian work on rocketry had reached the point at which manned space flight was technically possible. Like their predecessors in the Paris of the 1780s they were concerned to prove that interstellar space was a benign environment for human beings. And their method for establishing this was just the same; suitable animals should be launched into space and their reactions to weightlessness and whatever hazards also existed out there monitored. The programme progressed from tests from which the animal could not be recovered to the final phases in which they were brought back to earth unharmed and psychologically normal, as far as could be ascertained. When Yuri Gagarin went into orbit in 1961, like Pilatre de Rosier in 1784, his path had been smoothed by the experiments in which animals served as instruments able to detect dangerous but previously unknown hazards in the alien environment.[22]

Altogether the Soviet Space Programme tested the extraterrestrial environment with probes in which a total of thirteen dogs served as the instruments to detect conditions above the atmosphere. The most famous was the three-year-old Husky mongrel called Laika (Figure 4). Not only was she the first 'dog in space', but the rocketry could not at that time accomplish a safe return to earth and her life was sacrificed as the life support system in Sputnik 2 ran out.

Laika had been a stray picked off the streets. She was trained for her role as a living instrument in an earth-bound version of the same type of small capsule in which she was eventually launched by a modified Soviet intercontinental missile rocket. The capsule was comfortably padded and the animal was supplied with sufficient food and water for the short life that remained to her in the experiment. Basic telemetry recorded her vital signs, such as pulse, blood pressure, and respiration. The system kept her alive for several days. During this time she showed no signs of bodily

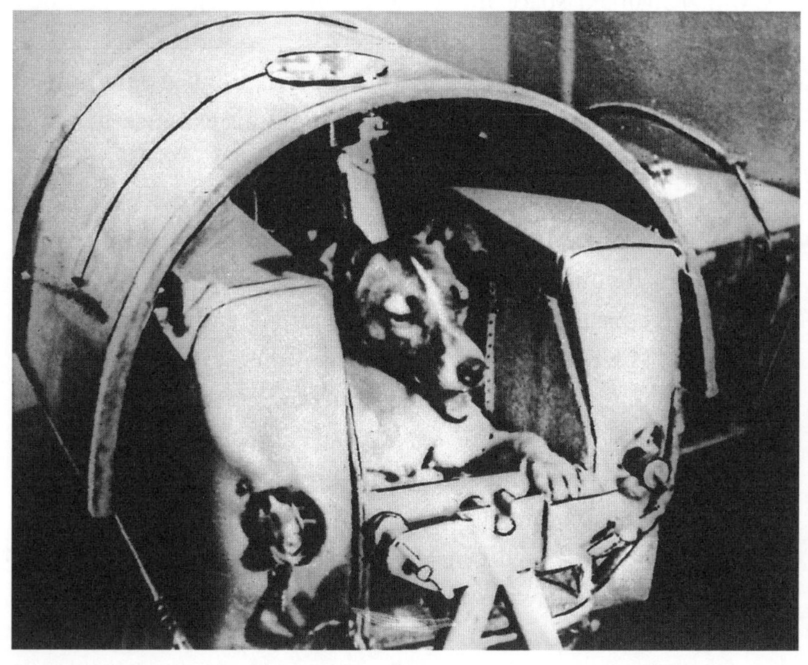

Figure 4. Laika the dog in her space capsule.

deterioration, though, of course there was no way that any psychological effects of the claustrophobic confinement in a tiny 'kennel' could have been known.

As Pavlov had caused a monument to be erected to the dogs who died in his studies of the mechanism of digestion, so Laika has a place in the memorial to the cosmonauts who had been lost during the space programme. The monument is in the aptly named 'Star City' near Moscow.

The aim of the programme was to send a human being into space and to bring the intrepid cosmonaut safely back to earth. The dog version of the trip was successfully accomplished in 1960. Belka and Streika spent a day in orbit and were safely brought back to earth. There were other living beings in the capsule as well, including plants. The results of the experiment were dramatically displayed by the fact that Streika later gave birth to six puppies. These dogs too have a permanent memorial. Their bodies are preserved in the Memorial Museum of Astronautics in Moscow.

The subsequent history of 'animals in space' involved the use of more and more species of animals, including chimpanzees by NASA. Ham the Astrochimp orbited the earth and returned safely in 1961. Even spiders and frogs were sent aloft. By this time the logic of the experimental programmers began to diverge from that which lay behind the Sputnik trials. In those trips the animals had been part of the equipment, sensitive detection devices tuned to pick up any malign influences that might exist in extraterrestrial environments. Later experiments were directed at a different research question—what effect would weightlessness and radiation have upon living creatures? The animals themselves became the target of experimental manipulations surrounded by inorganic instruments. The programme had moved from tests of the safety of space flight technology to exobiology.

Commentary

The examples of living instrumentation we have looked at in this chapter serve to show just how widespread is the use of organic instruments and how diverse are the projects in which they have played a part. The necessary causal links between the state of the environment to be detected and the induced state in the organic detector have for the most part not been life-threatening. Some of the space dogs died and indeed some of the astronauts perished in the interests of a mixture of the search for scientific knowledge and the bolstering of national pride. Cynics have noted the political advantages that some of these projects offered.

Both plants and animals have found uses as detectors. There are many non-scientific uses of organic detectors. The use of dogs to follow a trail or to pick out luggage in which drugs are concealed depends on the same principle of reliable causal influences from the target to the sensitive being acting as a detector.

Chapter 3
Measuring: The Bodies of Kings, Fossils, Lichen, and Pollen Grains

Among the most important pieces of equipment in the instrumentarium are devices which are used to *measure* various properties of material things and natural processes. Shape, size, weight and mass, speed of movement, the time a process takes to completion, temperature, and many other aspects of material things, substances, and systems need to be measured.

There are two fundamental aspects to measurement: the fixing of units so that standardized scales can be constructed, and the invention and construction of devices by means of which measurements can reliably be made. Living beings have been used in both roles. Until the past two centuries or so the official measures of length, area, and weight were related directly or indirectly to the shapes and sizes of organic beings. Most often these were the sizes of parts of the human body, sometimes the particular body of a king. Even barley corns played an important role in fixing standards of measurement. The body temperature of Fahrenheit's wife as well as his predilection for duodecimal numbering is incorporated in the temperature scale still in use in some parts of the world. Gradually though the living component has been deleted from the measurement standards in common use.

However, the use of the bodies of creatures both alive and dead as measuring instruments has remained an important feature of certain temporal measures. For example, for measuring the rate at which the climate has been changing, the use of fossils of animals, plants, and even bacteria, living or once living beings, remains at the heart of the scientific exploration of the past. Sometimes the best that can be done in measuring a sequence of events is to establish the order in which they happened. How long some process took, measured in years, say the development of flowering plants, or the dying out of a forest, may be very difficult, if not impossible to measure in any reliable way. Nevertheless a very good idea of the sequence of events involved can often be established. There are ordinal as well as cardinal numbered representations of the time ordering of sequences. In setting out an ordinal sequence how long each era lasted is not incorporated in the layout. All that matters is the order of the stages: first, second, third, and so forth. A sequence of events can also be represented by cardinal numbers, how many years elapsed at each phase of the story, a million, one hundred million, and so forth.

Establishing standard units

Large-scale units of time derived from astronomical observations have been used since antiquity. Time units derived from the cycles of human life have almost always been secondary measures, tying the vagaries of the human life patterns to astronomical sequences. The lifespan of human beings has never, so far as I can ascertain, been used as a standard. Lifetimes are scattered around three score years and ten, or rather they once were.

Devising small-scale units of time and the mechanisms to measure them has been more difficult. Successive days are never the same length. So the day becomes a fraction of the year and an hour a fraction of the day and so on.

Most kinds of measurements can take two forms. By choosing an object to serve as a standard, repeatable length, quantitative questions about the linear extent of material things can be answered to any desired degree of accuracy, depending on the choice of standard. Choosing a suitable unit, such as the length of the arm of the king from shoulder to fingertip, a numerical measure of the extension of an object can be obtained—so many ells, yards, metres, and so on—by repeatedly laying the standard measuring stick or rule along the object according to some suitable convention. However, if the units changed with every accession to the throne, no one king being exactly the size as another, chaos would result. So, in practice, just one royal body served to ground the metric, just as one woman's body temperature, that of Frau Fahrenheit, served to fix the scale inaugurated by her husband.

FIXING UNITS FOR TIME

It seems that units of time have always been derived from natural non-organic processes such as the movements of the Sun and Moon and, more loosely, the seasons. The irregularity of the human pulse made it unreliable as the source of a unit of time, though Galileo did refer to it in discussing an independent test for the regularity of the water clocks he used in his experiments on the time it takes a ball to roll down an inclined path. Other organic cycles suffer from the same defect, but have been tied 'mystically' into the cycles of the Sun and the Moon.

FIXING UNITS FOR SPACE

All that is needed for a measure of length is a decision to adopt a particular object as the metrical standard. Then by reusing that object or a representation of it according to certain rules a measurement of length can be obtained. A human body will do very well as such a standard, at least locally. A remarkable example is the case of the unit 'the smoot'. The story is related by Robert Tavernor in *Smoot's Ear: The Measure of Humanity*.[1] Candidates for one of the student societies at the Massachusetts Institute of Technology must undergo a physically rigorous initiation. In 1958 Oliver Smoot's initiation involved him being chosen to provide a unit of length with which to measure a bridge across the Charles River, connecting Cambridge with Boston. Smoot was required to lay himself out successively from one end of the bridge to the other. The length of the bridge was '364.4 smoots ± an ear'. This remarkable event is commemorated by inscriptions that can be seen on the bridge to this day. Perhaps inspired by his undergraduate experience as a unit of length Oliver Smoot went on to become the President of the International Organization for Standardization, of among other things standardizing measuring units of all kinds.

Medieval kings made their mark on the measurement systems of Europe both by lending parts of their bodies as the basis of metric units and by systematizing the units they had inherited from antiquity. As Tavernor points out, the human body served not only as a source of units for arithmetical measurements but also for proportions.

The 'foot' or *pes* was established as two distinct units in the Roman Empire. There were six *pes* in the Western Empire to five *pes Philataereus* in the Eastern Roman Empire. Both units were divided into 16 fingers or *digits*. That the English *foot* had its origin in the Roman *pes* is suggested by the fact that they were very nearly equal. The English *inch* derives from another Roman unit, the subdivision of the *pes* into 12 *uncias* or *inches*. Of course, human feet differ greatly in length. It is sometimes said the modern 12-inch 'foot' was derived from the feet of the English monarch Edward I. In fact the Normans brought the 12-inch foot with them after the Conquest. The *yard* was fixed in the twelfth century by Henry I, as 3 feet. The story that it was defined as the distance from the tip of the King's nose to the end of the thumb of his outstretched arm is another royal fabrication. For centuries this unit competed with the 'ell', from 'elbow'. The ell was widely used as a measure of cloth, and was related to double the length of a doubled forearm. It is easy to see how this unit came to be used in the clothing trade: when as a child I went shopping with my mother, the haberdasher would fold a length of cloth around the elbow and hand along the forearm. The inch, the basic unit which underlies both foot and yard, was fixed by the Normans as the length of 3 barleycorns.

Nevertheless the use of outstretched arms in defining a larger unit of length can certainly be found in antiquity. In the Ashmolean Museum in Oxford there is a bas-relief displaying both a *fathom*, the 6-feet spread of the extended arms and hands, and a *foot*.[2]

The height of a horse is measured in hands. Just as feet are very variable in length so are outstretched hands. Eventually the unit was standardized at 4 inches, a rather small hand, but then jockeys are on the small side.

A mile is the distance covered in a thousand paces (Latin *mille passi*) by a Roman Legionnaire on the march. The distance is 1,614 yards. The English mile of 1,760 yards does not seem to be a simple derivative of the Roman measure. The military measure also gave rise to the *stade*, 125 *passi* or paces.

The common measure of area prior to the establishment of the hectare, the *acre*, relates to the Roman measure of 12 feet, the *pertica*, from which came the obsolete *perch*. The English system for land measurement involved *rods*, *poles*, and perches, all of which are equivalent to 5.5 yards. Another living creature gets into the story here, the ox. The old English measure of a *furlong* was a contraction of 'furrow long' while the acre began as the rough and ready and highly variable area that a pair of oxen could plough in a day. Eventually it was formally defined as 160 square rods, and so 4,840 square yards.[3]

All this is now ancient history. Almost everywhere the metric system reigns supreme, owing nothing to the dimensions of human or animal bodies nor to the dimensions of cereal grains as the source for its standard units. In 1795 the metre was set at one ten-millionth of a quarter of the diameter of the earth, as it had been measured at the Paris meridian. The litre was fixed as one thousandth of a volume of one cubic metre and the kilogram as the weight of a litre of pure water. Smaller and larger units were created as required by factors of 10. The metric system was adopted in Germany in 1868 and by a consortium of nations a few years later. It was not adopted in the United Kingdom until the mid-1960s, even then in a piecemeal fashion. The old Imperial measures are still in use in a few countries, including the United States. It is perhaps ironical that the shadow of royalty is cast over the system of measurement of so staunchly republican a nation.

FIXING UNITS OF TEMPERATURE

Fahrenheit created units of temperature by deciding on two standard levels of warmth and coldness and then dividing the temperature spread between these naturally fixed points into equal parts. He set up the scale according to duodecimal principles, preferring ratios expressible in units of 12, or some submultiple thereof.

Daniel Gabriel Fahrenheit was born in Gdansk on 24 May 1686 and died at The Hague on 16 September 1736. He had a distinguished career as an instrument maker. The popularity of his mercury-in-glass thermometers ensured the spread of the scale he had devised. From the point of view of the customary user of the Celsius temperature scale the strange numerology of the Fahrenheit scale running from 32° for the freezing point of water to the boiling point at 212° seems bizarre. However, his choice of defining situations had a kind of rationale, in relation to his interest in the practical importance of the measurement of temperature. Zero degrees was fixed inorganically by reference to a mixture of ice and salt or ice and ammonium chloride, the lowest temperature conveniently produced in the laboratory. The temperature of the human body was important and conveniently taken with a mercury-in-glass thermometer of his own design. He determined that fixed point organically by measuring his wife's temperature in the armpit. Favouring a duodecimal system, in units of 12, his wife's body temperature could be defined as 96°, that is 8×12 units from 0. Setting the freezing point of water at 32° fits in as one-third of the temperature of the human body. Fahrenheit was an essentially practical man, and he had an eye for the commonest daily uses to which a thermometer might be put. Anyone can tell when water is on the boil. However, the temperature at which this occurs was known to differ according to the atmospheric pressure. Happily for the duodecimalist there is a difference of 180°F between the freezing point of water and its boiling point at the surface of the Earth. This has 12 as one of its factors.

Fahrenheit's thermometers were very popular and he set up a thriving business for their manufacture and distribution. Of course, in time there were demands for instruments that would measure temperatures far above the boiling point of water. He went on to construct thermometers that could measure temperatures up to about 600°F.

In time the centigrade scale displaced his duodecimal scheme in the sciences, and has gradually spread to most situations in which it is important to measure temperature. In 1948 an international commission agreed to rename the centigrade scale after Anders Celcius (1701–44). Aiming at a universal scale he tested the reliability of the freezing and boiling points of water to define a fixed temperature standard at sea level quite independent of any tenuous links to the attributes of living beings. His scale ran from 100°C for freezing to 0°C for boiling water. This odd layout was reversed by the great biologist Charles Linnaeus. And so it has remained to the present day.

Measuring devices

DEVICES FOR MEASURING TIME

The point of constructing clocks is to make available a sequence of repetitions of a standard unit of elapsed time to measure time spans. The clock maker must find some process that seems to repeat itself more or less exactly and then settle on that as the basis of the time measurements by definition. Measures of time are always relative to choice of repetitive process to which a definition of a unit of measurement can be attached. However, it may sometimes be of equal importance to be able to arrange things in a certain order, say in greater or lesser volume. Objects can be arranged in order of length and so on. The ordinal principle would simply be something repeatable but asymmetric, such as 'larger than' or 'longer than'. How much larger or longer cannot be represented on an ordinal scale. Solid substances can be arranged in a table of relative hardness, using Moh's scale—which stuff can be scratched by which. Time lapse can be determined quantitatively by clocks, but the time order of events, such as the laying down of geological strata, is also a matter of great scientific interest, even though at first the span of the geological ages in terms of a repeatable standard such as the year, could not be determined. In this chapter we will find both ordinal clocks and clocks capable of giving a measurement of the time elapsed between the events arranged on an ordinal scale.

Let us look at some mechanical devices for time measurement, a clock or clocks, which will set the scene for the story of organic measuring devices.

Mechanical devices

The sundial has obvious drawbacks as a clock. There are seasonal differences in the length of the day and fine measurements are difficult to make. The water clock, or clepsydra, developed in Greece and Egypt in the pre-Christian era, depended on the uniformity of the flow of water through a small hole into a graduated vessel. Such a device could be used at any time of the year and during the hours of night. When the science of mechanics was studied later in the Middle Ages and the Renaissance, accurate measures of time were needed. Mechanical clocks had been invented, and some were of considerable accuracy, such as the clock of Richard of Wallingford.[4] This clock was based on refinements of the type of machinery used in a flour mill, and though adequate for domestic purposes, its scientific use was only relevant to astronomy.

However, the clepsydra was in use for scientific purposes. It could be designed to measure fairly accurately quite short times. At the root of any clock mechanism there must be an orderly process of recurring events, a sequence that could be defined as 'regular'. The movement of the Sun across the sky, the dripping of a restricted stream of water, and the tick-tock of a mechanical escapement are all such sequences. However, none were really adequate for the kind of measurements needed in such investigations as Galileo's study of the mechanics of falling bodies, as they rolled down an inclined beam. Each clepsydra had its peculiarities. The need to maintain a temporal standard for repeated experiments done by other investigators could not be satisfied by building more clepsydras. It would have been necessary to settle on a particular one to provide the standard sequence.

Here is Galileo's description of the timing device he used in his experiments.

> For the measurement of time, we employed a large vessel of water placed in an elevated position; to the bottom of this vessel was soldered a pipe of small diameter giving a thin jet of water, which we collected in a small glass during the time of each descent [of the ball down the groove on the sloping board], whether for the whole length of the channel or for part of its length; the water was collected and weighed, after each descent, on a very accurate balance; the differences and ratios of these weights gave us the differences and ratios of the times.[5]

But what was the standard against which the results from the clepsydra were checked? 'We repeated this experiment', says Galileo, 'more than once in order to measure the time with an accuracy that never exceeded one tenth of a pulse-beat.' Galileo's own heart played the role of the caesium of modern chronography, but it was an organic timepiece, and surely unreliable.

An ordinal measure of geological time: fossil clocks

Fossils are the petrified remains of living organisms, of plants, animals, and people. They played an essential role in the unravelling of the geological history of the Earth. Around the turn of the eighteenth century, the surveyor William Smith realized that he could use fossils to determine the time order of geological strata. The remains of animals and plants provided an ordinal scale of time, with which the geological structure of different parts of England could be made into a single coherent picture.

William Smith was born the eldest son of John and Anne Smith in the village of Churchill in Oxfordshire, on 23 March 1769. The family were of solid yeoman stock, owning their own farm. The Smith family were superior artisans, the father being both a blacksmith and a skilled mechanic. When his father died in 1777 William's mother continued to run the farm, eventually marrying again. William and his brother were taken in by his uncle and brought up on a dairy farm. The elementary schooling he received was not enough to hold his attention. His interest in geology was sparked by the realization of the nature of the 'pound stones' used in the dairy to weigh the butter. These were domed symmetrical objects, about 10 cm in diameter and segmented into five sections. They were found scattered in the fields around Churchill where they can be picked up to this day. William realized that they were the remains of sea creatures, in fact of sea urchins. The fields around the farm must once have been covered by the sea.[6]

For several years William managed to combine his work on the farm with intensive private study of geometry and the basic principles of surveying. At the age of 18 he was employed by a local Cotswold man, Edward Webb, to survey the parish of Churchill. Evidently things went well and William moved in with the Webb family. Edward Webb was an excellent mathematician and an accomplished instrument maker. In 1791 Webb sent William to undertake a survey of an estate in Somerset, which included a number of coalmines. He was able to observe the strata revealed by the mining operations. Having no means for measuring the age of the rocks in successive strata he developed an ordinal measure using fossils characteristic of each stratum as his time markers. His work came to the notice of Alexander Rennie, chief engineer to the Somerset Coal Canal Committee, and in 1793 Smith began to work on the surveying aspects of the construction of canals.

At this time a great network of canals was being built across England to facilitate the movement of heavy goods, above all coal. His travels in surveying canal-building projects gave him an opportunity to observe the layers or strata of the rocks through which the canals were being dug. In 1794 he was picked by the Committee to be one of a party to tour a large part of England to observe the work of other canal builders. It was during the inspections of works in other regions of the island that he got the idea of making a map of the subterranean levels of England, guided by the fossils to be found in them.

At first sight Smith's method of determining the relative age of rock layers seems to be open to a fatal objection. A start must be made somewhere. From a study of a particular cross section of layered strata at some particular place a catalogue of fossils can be made. Then strata in other locations can be arranged in order by reference to the fossil record in the original catalogue. There seems to be no independent way of finding the relative age of either the rocks or the fossils. Smith broke the circle, so to speak, by adopting the principle that overall, the uppermost strata were the youngest and the lowermost the oldest of the geological formations. The value of this move exploits a distinction well known in philosophy between truth as correspondence, where a statement is judged according to its fit with independent material evidence, and truth as coherence, where a statement is judged according to the way it fits in with a larger body of other statements, conjectures, and so on, the whole of which makes sense of what has been observed. Both conceptions of truth have their part to play in the sciences. Dating by reference to the decay of radioactive substances in the rocks eventually provided a 'correspondence' cross-check on Smith's coherence principle.

The fossils too showed a pattern of differences that fitted well with their locations in the levels of strata. The lowest or oldest strata had no fossils, while the successive layers displayed fossils of creatures of greater and greater complexity.

For the next few years William Smith worked as a surveyor for the Somersetshire Coal Canal Company. During this time he accumulated more and more examples of the link between stratigraphy and palaeontology. An unrivalled observer of those anomalies that are so revealing in science, he noticed that the strata sloped gradually towards the east. Did this mean that the same pattern would be found throughout England? His travels for the company and later on his own account gave him the opportunity to test his 'Principle of Faunal Succession' as he called it. Did the fauna, the animal remains preserved as fossils, display the same relation to the order of geological strata everywhere? Indeed, they did.

However, in 1799 Smith's life took another turn. He left his job with the Somersetshire Coal Canal Committee. His nephew and biographer, John Phillips, offers no explanation for the ending of this relationship. However, freed from the necessities of a demanding daily job Smith was free to roam about England collecting material for his masterwork, a geological map of England, first published in 1815. He had made his home in London and had been earning a living partly by the sale of his hand-coloured maps of the geology of various parts of England and partly by working as a self-employed surveyor. The exceptionally wet weather of 1799 had forced him to develop his skills in drainage, and this made up the bulk of his work.

Copyright laws hardly existed in those days and his maps were extensively plagiarized, and sold below the prices he had been asking. In debt and eventually bankrupt he was committed to the Kings Bench Debtor's prison in London in 1817. Released a year later, he found bailiffs in possession of the house. They removed every last thing he had left.

Over the following years he worked as a surveyor for various employers. During his itinerant wanderings after leaving his job at the coal company he had met many of the leading men of the time. Everyone of consequence enjoyed country estates and surveyors were in demand. Though his work was well known through his maps, official recognition did not come until Sir John Johnston, an amateur geologist, took up his cause. In 1831 the Geological Society of London conferred on him the Wollaston Medal. From then on his career prospered in all sorts of ways. As the acknowledged 'Father of English Geology' he received an honorary doctorate from Trinity College, Dublin. Various official posts followed. That the son of a humble country blacksmith should be a member of the commission to select the stone for the Palace of Westminster, the new Houses of Parliament, must have struck him as a pleasant irony. He died on 28 August 1839 in Northampton.

Another ordinal measure of geological time: molecular fossils

Thinking of fossils brings to mind chunks of rock with interesting structures and patterns clearly the impress of living beings. These are the sorts of things that people can pick up on the beach at Lyme Regis or chip out of the wall of a quarry. However, bacteria and other simple organisms, such as the archaea, can also become 'fossilized' but in an indirect way. Molecular fossils are the remains of the life processes of fragile beings which have long vanished themselves, leaving little or no trace of their bodily forms. However, their molecular residues are of the utmost importance in measuring the age of oil deposits, for example. Each species of micro-organism leaves its characteristic molecular 'signature' so to speak. There are two dimensions of difference that have proved scientifically important. Some bacteria metabolize oxygen (aerobic species), while others metabolize elements such as sulphur, some thriving on a diet of hydrocarbons. These are the anaerobic species. Some micro-organisms flourish in moderate temperatures. Others, for example the thermophiles, lovers of heat, can live and reproduce in temperatures up to 150°C.

Some years ago the cosmologist Thomas Gold (1920–2004) presented what he declared to be a radically new theory of the origin of subterranean oil deposits. There is no doubt that there are molecules in oil, coal, and natural gas that seem to indicate an origin as the remains of plants that once flourished on the Earth and lived by photosynthesis. It was claimed by the Russian geologist N. A. Kudryavtsev in the early 1950s that rather than resulting from chemical transformations of buried plant material, coal and oil deposits are of inorganic origin, produced in the deep mantle of the Earth, the rocky layer above the iron core. The molecular evidence that suggests a surface organic origin is the result of contamination by molecular fossils. Some have come from surface plants that have been buried by geological processes. But there are others also to be found in oil and coal, so the theory proposes, that are indicative of anaerobic types of organisms that do not depend on oxygen for their life support. They show, it is alleged, that there are life forms deep in the Earth that feed on gaseous hydrocarbons leaking up from great depths, far below the regions in which we now find oil and gas deposits. Gold proposed a very similar 'abiogenic' theory of the origin of the hydrocarbon deposits such as oil and coal. Basic hydrocarbons, such as natural gas, are, he declared, the products of inorganic processes deep within the Earth. The organisms of the 'deep, hot biosphere' feed on this material, transforming it into oil and coal.

Thomas Gold grew up in the unstable conditions of the rise of the Nazi regime in Germany and the disruptions of the Second World War. Perhaps this accounts in part for his readiness to propose 'daring' hypotheses. He was born in 1920 in Vienna, but when he was ten the family moved to Berlin and then to England. Thomas read both physics and engineering at Cambridge. Interned early in the war as an enemy alien, he was quickly released to become a leading member of the team developing naval radar. Though the 'steady state' theory of the Universe to which he contributed with Herman Bondi and Fred Hoyle eventually lost favour, his theory that pulsars are rotating neutron stars has proved to be correct. As he said of himself, 'In choosing a hypothesis there is no virtue in being timid. I clearly would have been burned at the stake in another age.'[7] Despite the reservations one might have about the theory of an abiogenic origin of the hydrocarbons and the deceptive role played by the biological remains that form this version of an ordinal bio-clock, there is just enough supporting evidence from biological and geological research to make one uneasy in dismissing the ideas of a man who has often been proven right in the end.

> How much of the biological imprint of material in the [hydrocarbon] sediments is due to surface life and how much to life at depth? Do the biological molecules of [in] petroleum and coal indicate now merely the additions from the deep biosphere to material of primordial [inorganic] origin, rather than indicate a biological origin of the bulk of the substances themselves?[8]

He summed up the theory in a neat aphorism: 'Hydrocarbons are not biology reworked by geology (as the traditional view would hold) but geology reworked by biology.'

The key to deciding between the two theories, surface origin and depth contamination, should be to consult the molecular fossils that have been used as markers of the age and origin of such subterranean substances as coal and oil. The standard dating system is based on the presence of molecular fossils which seemed to indicate that they had come from surface living organisms which had depended on photosynthesis, the carboniferous theory. If hydrocarbons in the form of natural gas had been seeping up from below the biosphere it should also be contaminated with molecular fossils indicative of the existence of another strain of micro-organisms that lived at even greater depths. These beings would not have been dependent on photosynthesis for their life energy.

The molecular 'signature' of an origin of hydrocarbons from life forms is most clearly legible in *hopanoids*, which are the molecular remains of membranes of living cells. They are derivable only from prokaryotes, single-celled organisms that do not have a clearly demarcated nucleus to which genetic material is confined. They include the bacteria and the archaea, the two main groups of prokaryotes.[9] Hopanoids differ in hydrogen content, being higher in those derived from organisms which live by photosynthesis and lower in those which live by making use of sulphur or hydrocarbons. So if some hopanoid molecules with low and some with high hydrogen content are found in oil, this suggests not only that the oil has been in contact with the remains of a surface region of organic life, but also that it must have been in contact with the organisms of a 'deep hot biosphere'—a region where extremeophile life forms capable of surviving under conditions of great temperature and pressure would exist. They would have something in common with the sulphur-consuming thermophiles of the genus *Sulfolobos*, found in certain volcanic vents under oceans and the boiling lakes of Yellowstone Park. If greater depth is correlated with greater antiquity the hopanoids can serve as the basis of a very crude but effective ordinal geological clock.

The most startling hypothesis that appears in Gold's theory is his claim that the biosphere, the region in which living things exist, extends much deeper into the Earth than had hitherto been imagined. Indeed more and more evidence has shown that in this he was almost certainly right. In a recent paper in the journal *Nature*, R. John Parkes remarks,

> large bacterial populations have been found at considerable depths (up to a few kilometres) in a range of different subsurface environments . . . There are even some locations where bacterial activity and populations increase in deeper zones and where

> geospheric [as opposed to atmospheric] processes may be involved in fuelling a bacterial biosphere to much greater depths than we had previously thought possible.[10]

The oldest fossil bacteria have been found by Roger Summons in the rocks of Western Australia.[11] They are estimated to be about 2.5 billion years old, and can be seen as patterns in rocks, much as one sees the skeletons of dinosaurs. However, the dating comes from the hydrocarbon derivatives they contain that are indicative of cyanobacteria. These were photosynthesizing organisms that transformed the Earth's atmosphere by feeding on methane and producing oxygen. Even older is the evidence for life claimed by the Danish geologist Minik Rosing to prove the existence of living beings in Greenland rocks from 3.7 billion years ago.[12] However, doubts have been cast on this claim. Rosing has based his claim on the discovery that the ratio of the isotopes carbon-12 and carbon-13 in the minute carbon grains found in some Greenland rocks is typical of the remains of once-living organisms. However, the same ratio between these carbon isotopes can also be the result of inorganic processes in the formation of such granules.[13] The consensus among geologists seems to be that the alleged evidence for traces of early life in very ancient rocks is not definite. By itself it does not show that the 'indicative' ratio could not have had a purely inorganic origin.

What has this to do with the measurement of time? In the theory that oil deposits are the result of surface processes, estimates of the age of the subterranean deposits must be limited by the geological record of the sequence of kinds of plants that flourished on the Earth's surface. The Gold hypothesis subverts this thesis at least in part. If he is even half right the ordinal dating of the creation of oil and coal cannot be tied neatly to the carboniferous era. No one knows when and how the proposed deep anaerobic bacterial domain was established, and whether it flourishes there still.

Dating climate change

An important use for a metrical system of time measurement, going beyond the ordinal scales used in fossil clocks, is to date important events in the past history of the Earth. The idea of investigating fossil plants to trace the history of certain geological processes had occurred to Charles Lyell in the first half of the nineteenth century, just about the time that William Smith was using the fossil remains of animals, fish, and insects for the same purpose. In 1845 on a visit to North America Lyell studied the way vegetable matter accumulated in alluvial sediments as the precursor of coal. These remains ought to provide a record of the plants that had been growing near to the wetlands in which their remains were deposited. This was not the only situation and these were not the only kinds of plant remains from which climatic inferences could be made on the basis of vegetable data. Smith's chronology was ordinal, an attempt to arrange geological strata in temporal order. But could the length of time that each geological era had lasted be measured? Would it be possible to find a method for determining when a certain climate change had occurred in terms of standard units like years? We now have various forms of dating based on the principles of radioactive decay—but before then, efforts at dating were vigorously pursued.

Trees, grasses, and many other plants produce great quantities of pollen, which are spread far and wide by the wind. Pollen grains are small and physically tough particles, the male gametes of sexually reproducing plants. Plant spores differ in shape, so that microscopic examination fairly easily reveals the parent plant. Only a tiny proportion of these particles fulfil their biological destiny of fertilizing the female gametes, the ova in the flowers of their appropriate species. The rest for the most part settle on the ground and some find their way into bogs, marshes, and lakes. In suitable circumstances pollen grains remain intact for many thousands of years. They have the potential to reveal the past history of the plants in areas where they had been captured and preserved.

One such preservative environment is provided by the peat bogs of Ireland. Another is a similar terrain in Sweden where the use of pollen to reveal the past botanical history of a region began in the nineteenth century with the work of F. Trybom (1888). Using samples from the sediments in Swedish lakes he tried to form an idea of the botanical history of the neighbourhood. He coined the term 'index fossils' for the pollen grains he found in the deposits on the floors of lakes. It was not long before it was realized that the 'plant-scape' of a region was a good indicator of the climate when the pollen was deposited, and so by comparison with the flora at that place in other eras and with the present, of climate change. The first detailed analysis of the fossil pollen in Swedish bogs was published in 1916 by Lennart von Post. His comparative study of pollen grains could be developed not only into providing an ordinal map of events such as changes in climate but a way of dating them as well by counting annual sedimentary deposits.

Lennart von Post was born on 16 June 1884. After university studies in botany and geology and many years of field studies, he was appointed to a professorship at Stockholm University in 1929. He was the first to publish analyses of fossil pollens. He was evidently a man of great energy, expecting the same dedication to the job from his assistants as he himself displayed. He would always be the first to arrive at the site of a field study, dressed in a voluminous overcoat, all the necessary equipment hanging from various points about his person. He retired in 1950, dying shortly thereafter on 11 January 1951.[14]

Since his time this branch of geology has acquired its own name, 'palynology', the study of fossil pollen grains, spores, and other minute particles preserved from the plant life of the past. The invention of the name is credited to H. A. Hyde and R. F. Adams.[15] It is derived from the Greek word '*pale*', meaning a fine dust. This word is related to '*polle*', the Latin word for flour.

In order for pollen studies to be relevant to assessments of climate change there must be a strong link between pollen profiles in the peat and the plant life of the time at which the pollen was produced. Plant species are highly characteristic of climatic conditions. By identifying the prevailing species at a certain time, pretty robust inferences about the climate of that era can be made. Research into this matter has shown that while the fit is not perfect, nevertheless the types and abundance of species is reflected in the pollen frequency of each level in the bog.[16]

Pollen grains fall onto the surface of the peat bog in the spring and early summer, and are covered by the remains of the plants that each autumn form the next layer of peat. Gradually these become compacted into the almost coal-like material that used to warm the houses of Irish country folk.

The formation of peat is a slow process. A depth of a metre may contain the potted history of thousands of years of climate change. It is even possible to identify the ash or 'tephra' of volcanic eruptions thousands of miles distant as the particles are borne on winds around the globe and settle onto the surface of the peat bog.

Determining the time spans of the climatic periods represented by the types of pollen in the peat requires physical measurements, the familiar technique of carbon dating. From the perspective of this book such methods lie outside our remit. Ash from a known eruption can also be used to date layers in the vertical cross section of the peat. Prior to the use of carbon dating the peat bogs gave good data on the *order* of climatic events, even if the exact time at which they had occurred was not precise. By counting layers and assuming them to be roughly annual, a reasonable estimate of age is possible.

Pollen grains differ greatly in size. At first sight it might seem as if the larger grains should stay in the upper layers of the bog while the smaller grains would sink down to lower levels through the interstices of the fibrous vegetable matter. In fact this does not happen, so the pollen profile of a particular level is a fair sample of the pollen deposits at the time at which this layer was laid down.

The basis of the pollen detection method is the very distinctive forms taken by the pollen grains of different plant species. For example, Hyde and Adams, in *An Atlas of Airborne Pollen Grains*, illustrate 103 types of pollen grains distinctive in shape and size of the species of plant from which they come (Figure 5).

In order to carry out an accurate assay of the pollen in a sample of peat much of the irrelevant detritus of ages must be purged away. Samples are treated chemically, and then dried and prepared for microscopic examination. The outer surface of the pollen grains is resistant to most chemical reagents, so the cleaning up of the samples leaves the grains intact. Pollen frequency is the standardized measure of pollen density, the number of grains in a cubic centimetre of peat prepared in a standard way.

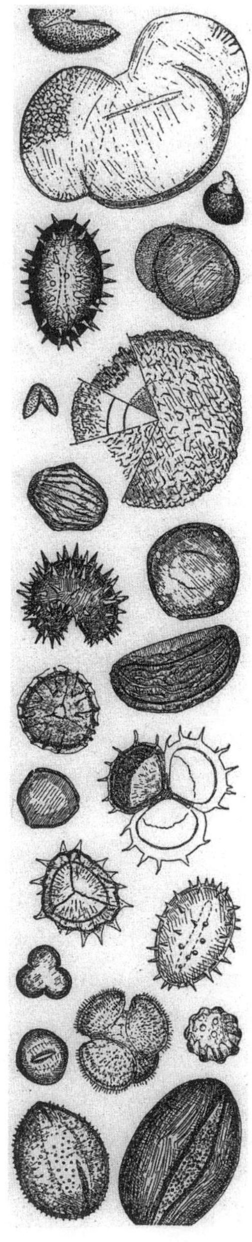

Figure 5. Shapes of pollen grains.

The 'equipment' for detecting climate change consists of the bog and its accumulated and hierarchically distributed pollen grains. However, the layer on layer of annual depositions of pollen and other remains can themselves be the measures of the time at which the layers were deposited, if the boundaries between the annual depositions are clearly detectable, just as we can detect the sequence of growth rings in the cross section of a tree trunk. This is rarely perfectly preserved in the peat bogs of Ireland.

Not only do peat bogs preserve the history of climate in the form of pollens, but so do lakes. The most striking contemporary research based on this phenomenon is centred on Lake Van in Turkey. The lake has certain unique characteristics that make it particularly valuable as a research tool. Since the rate of deposition of sediment on the lake floor is known, it serves not only as an instrument for detecting climate changes but also as a measuring device for determining when such changes occurred. There is no more need for radiocarbon techniques than there is when we determine the age of a tree by counting the rings.

Thomas Litt, and his team, from the Department of Palaeontology at the University of Bonn have studied drill cores taken from sediments at the bottom of Lake Van. The cores preserve the order of the deposition of the annual 'harvest' of plant remains, but these have become highly compressed. The deposition of sediment in Lake Van differs between summer and winter. In summer the sediment is mostly lime mixed with the pollen that drifts down from the surface of the lake. It is whitish in colour. However, in winter the sediment is mostly clay, brownish in colour. Like tree rings, the sediment layers are clearly differentiated year by year. Lake Van is not only a device for detecting climate change through the differences in types of pollen in the sedimental layers, but can also serve as a measuring instrument. The deposit of layers of sediment is hundreds of metres thick and must be the result of hundreds of thousands of years of repetitions of this process, summer and winter. Each layer may be no more than 0.05 mm in thickness.

Since the lake is about 400 m in depth, the deposition of the distinctive annual layers is not disturbed by the effects of surface weather. At that depth the water is quite still. According to Thomas Litt, 'there are about 20,000 annual strata to 10 metres [of sediment]. We presume that the bottom of Lake Van stores the climate history of the last 800,000 years', since the sediments are 400 m deep in places.[17] The cores are so compressed that they reveal their treasure of data only under high magnification.

Lichenometry

Lichenometry is the use of lichens to date rocks. The method was developed by Roland Bischel, an Austrian glaciologist. The method depends on the principle that the thallus or disc of the lichen as it grows on the surface of a rock steadily expands. Bischel measured the thalli of lichens on dated gravestones. It was easy to find the rate at which lichens develop by comparing the date on the gravestone and the diameter of a thallus found thereon with the size of the thallus and date of another gravestone. By choosing the largest lichens and assuming that their spores had germinated first on the gravestone, Bischel worked out a kind of living clock. He used his 'clock' to date the morraines, the ribs of stones, thrown up by the descent of glaciers.[18]

Lichenometry is still in use as a simple and practical method for dating rocks recently exposed to the atmosphere, where other measures of age are not easily applied.

Sidelights

These organic devices are part of the current instrumentarium of scientists working on palaeontology and other geological topics. However, behind these contemporary uses lies a powerful inorganic measuring instrument, the relative rates of decay of radioactive substances incorporated in the material which must be dated. The principle is very well known, and really quite simple. Radioactive substances decay at a rate that can be represented along a descending curve that relates a measure of residual radioactivity to time. The curve descends but takes a very long time to reach the *x*-axis of the graph along which time is measured. Each radioactive substance has a definite 'half-life', the time for half the atoms in the sample to have decayed. In the course of this decay process the radioactive element, say radium, is converted into lead. By estimating the proportion of the original radioactive atoms yet to decay in relation to the quantity of lead that has been produced, physicists can work out the age of the rocks, that is, work back to the time when the radioactive material was first incorporated. This gives an independent check on the estimates that have been arrived at by the use of biological material.

DEVICES FOR MEASURING TEMPERATURE

The midge thermometer

Detection devices and many measuring instruments depend on a reliable causal link between some natural phenomenon and a property of the instrument. To detect the presence of something of interest in the environment a bare causal link between the world and the instrument is all that is required, provided that the link is more or less immune to environmental conditions other than that which is the matter in hand. The efficacy of the ordinary glass/liquid thermometer as a measuring instrument depends on the fact that the column of liquid expands or contracts in a reasonably regular way with increases or decreases in the temperature of the object of interest, for example a human body or the ambient air. In these cases the causal relation is active directly and with almost no delay between the cause and its effect on the state of the instrument. In common use the amount that the thermometer warms up slightly, thus lowering the temperature of what it is measuring, is so small as to be insignificant. Moreover, so long as there is a reliable causal chain between the state of the object to be measured and the subsequent state of the instrument there is no reason why millennia need not have elapsed between the environmental cause and its observable effect. This is the case with the midge thermometer. By examining fossil midges in the here and now biopalaeontologists can read off the temperature of the environment millennia in the past. Here is how it works.[19]

By 1895 estimates of temperatures in the past were being made by studying the distribution of the remains of chironomids, non-biting midges, in lake sediments. A metric was created by the use of pollen dating, or palynology, as described earlier in this chapter. Though the time estimations were quite crude, nevertheless a sequence of midge fossils that covered the whole Holocene period, that is, the past 10,000 years, had been developed. The principle behind the use of midge remains is simple. The prevalence of different species of midges is dependent on quite small differences in the temperature of the surfaces of lake water, affecting the larval stage, and the ambient air, affecting the adult stage of these creatures. Each species has an optimum temperature at which to flourish. So a 'temperature calibration set' of chironomid species can be defined.

Recent work by S. J. Brooks and colleagues has pushed back the limit of temperature readings to the late glacial era, about 12,000 years ago. The technique has steadily improved. More careful taxonomy has refined the calibration sets by identifying different species with different temperature optima. The calibrations themselves have been tested by running a contemporary census of chironomid species prevalence. Using the known temperatures over recent years and correlating these against the proposed calibration sets of the prevalence of midge species the 'thermometer' can be checked. Generally the fit has been good.

Most of the work of inferring historical temperatures has been based on studies of lakes in northern Europe. Brooks notes that midge populations and hence midge fossil remains in lake sediments are also influenced by the trees and plants that border the lake. During the final stage of the last Ice Age vegetation returned only slowly to the deglaciated regions of Europe so temperature rather than plant life would have had the greatest impact on chironomid populations, and thus on the fossil record. In the Holocene, reforestation and the spread of lakeside plants would also have had an effect on the midge population and distribution. Brooks reports work on creating calibration sets of fossil chironomid by picking lakes not affected by these post-glacial influences, such as by being above the treeline. As anyone who has hiked in Scandinavia knows, the treeline is not the limit for midges of either the chironomid or the biting varieties.

An obvious condition for treating the midge thermometer as a reliable instrument is that no random and disturbing influences should have come to bear on the fossilized midge corpses over the time of their preservation. The sediments in deep, cold lakes are remarkably free of disturbances.

The Greenland molluscs thermometer

Plants and their products are one source of information about climate change. Animals can also serve as a source of evidence. It is pretty clear that the presence of woolly mammoths in a region suggests that it was colder than the tropics in our era where the almost hairless modern elephants now live. There are more subtle ways in which animals can be used to register climate change than in the thickness of their coats.

Two geological periods are of special interest to scientists attempting to get an understanding of global warming. The Eemian was the penultimate interglacial warm spell, dated about 120,000 years ago. At the start of the present era, the Holocene, there was a short warm time, the Climatic Optimum, about 10,000 years ago. Knowledge of what happened in both periods is important because we think we know some of the changes in the distribution of plants and animals that preceded each and what happened afterwards. If we could detect similar changes occurring now we would have some data for making predictions about how the current warming will turn out, particularly for the organic world, including ourselves.

By cataloguing the kinds of molluscs that now inhabit the colder Eastern side and the warmer Western shores of Greenland and comparing them with the distribution of fossils of their ancestral forms and relating them to a timescale derived from radiocarbon dating one can use the molluscs not only as detector but, once calibrated, as measuring instruments too. The ranges of their distributions year on year, retreating from a cooling north and returning as it warms, gives a kind of calendar which tells us roughly where we are on some more extensive timescale of climate change. This is not a purely organic measuring device, since just as Roland Bischel calibrated lichens by the dates engraved on gravestones, so the molluscs must be calibrated by an inorganic device, the decay of radioactive carbon.[20]

Some species that once lived there are now absent from the Eastern shore, but are among those now found only in the West. The temperature ranges in which these creatures can live are known from studies of their modern descendants. It is an easy calculation to work out how much warmer the Eastern shore was in the Eemian period. From the fossil species that flourished then, the climate in the summer was probably about 5°C warmer than it is at present. By similar reasoning the summer temperature in the Climatic Optimum was also higher but only by about 2°C. Between 8,000 and 4,000 years ago the 'thermophilous' or 'warm-loving' molluscs died out on the shores of Greenland.

The work of excavating the fossil species and collecting specimens of living species was organized and largely carried out by Glenn Jones of the National Oceanic Sciences centre at Woods Hole. This involved demanding journeys, mostly on foot, through a difficult terrain.

One species, *Zirphaea crispata*, found only in fossil form in Greenland, currently lives elsewhere in water around 9°C. *Panopaea norvegica*, a species happy in water of 3°C, arrived on the Greenland shore before *Zirphaea*. The time between these arrivals gives a fair estimate of the time for the temperature of the coastal waters to rise by 6°C. In geological terms this was quite short. To clinch the matter the researchers needed to confirm that *Panopaea* came to the Greenland shore before *Zirphaea*.

Commentary

In this chapter the material has been organized around two very different ways in which living things have been involved in measurement. Various aspects of human bodies, such as temperature, length of stride, size of forearm, were used to determine one important feature of measurement, the essential units for quantitative studies. Once these units were transformed into something material and stable, their organic origin ceased to be important. The use of the forearm to measure cloth is a rough-and-ready device to measure the length of a bolt of cloth in 'ells'. Graduated rules, standard lengths of chain, light years, and so on have been the predominant instruments for spatial measurements and have almost always been material, whatever might have been the origin of the measurement units. Only in the case of the unfortunate Smoot did standard unit and measuring instrument coincide. On the other hand, measuring instruments for lapses of time have regularly been made out of living beings or their parts and remains. Pollen grains and bog and lake sediments are themselves instruments with which measurements are made. Thermometers too have used the remains of organic beings to measure changes through geological times. Fossil midges and ancient Greenland molluscs are some of the devices that one could call 'organic clocks'.

Chapter 4
Extending an Established Domain: Dogs, Horses, Frogs, and Trees

Scientific research programmes have often been directed to a kind of exploration of a material system. The aim is to discover the structure of the system, its constituents, and the details of whatever processes occur in it. The long struggle to understand the system of the Sun and the planets began in antiquity and is still actively pursued today. Over the centuries new questions have arisen on the back of solutions to old ones, and new equipment and techniques have been invented and developed to answer them. What does the Solar System consist of? How and why do its component parts move the way they do?

In this chapter the material systems to be explored are the bodies of animals and plants. There are vital fluids in the bodies of living things: blood and sap. What is their role in life and how do they flow around the body? All animals breathe in air and exhale it in a somewhat changed state. What happens to the air in the lungs? At the time of the experimental programme described in this chapter the idea that plants breathed too was gaining ground. Much was simply unknown.

A great deal was written in antiquity about the mammalian blood system. Like much else inherited from ancient Greek scholars such as Aristotle and Galen the spirit of the Renaissance was sceptical of their authority in matters of the workings of the human body. The problem of the nature of blood and of the arterial and venous systems and of inhaled and exhaled air was at the forefront of scientific interest in the seventeenth and eighteenth centuries.

Rather than experiment on models or analogues of the mammalian venous and arterial system, the apparatus of choice was the living body itself, and more particularly the bodies of domestic animals. Dogs and horses predominated in the instrumentarium of the Oxford 'cognoscenti' and that of the great Cambridge physiologist Stephen Hales. During the same period the role of air in sustaining life came under experimental study. To try to answer these questions animals were incorporated into experimental apparatus, particularly trapped under bell jars where their struggles to breathe were carefully noted while the air was extracted by the use of the newly developed vacuum pumps. Similar experiments were carried out to try to understand the role of air in the life of plants.

Relatively few animals and plants were involved in the experimental programmes. The generalization of results beyond the laboratories of the experimenters depended on the principle at the root of most scientific thinking: one or a few studies of representative items from a population are enough to base assertions about any member of the population, provided those studied are typical. And of course a sceptical philosopher can ask, 'How do you know that the one you have been working with is typical of its kind?'—to which the working scientist can reply, 'Only time will tell!'

These studies ran parallel to similar research programmes in the physical sciences—hydraulics and pneumatics. Stephen Hales, one of the leading experimenters on the flow of blood, was consciously pursuing a research line parallel to that of Robert Boyle in his discoveries about pneumatics, the behaviour of gases in enclosed spaces. Boyle discovered that a change in pressure is linked with a change in volume of an enclosed gas. Hales made a carefully designed attempt to find a law relating the pressure and volume of quantities of blood enclosed in the system of veins and arteries. If there were such a law it would surely be a key principle in the hydraulics of blood. The study of hydraulics, the motion and pressure patterns of liquids in tubes, served as an inspiration for Hales's studies of plant circulation. Here, too, his experiments were aimed at finding quantitative laws relating pressure and rate of flow of sap in the trunks, roots, and leaves of plants.

Blood: the venous and arterial system of mammals

The living experimental apparatus to be described in this section was developed by a group of English scientists, 'natural philosophers', in the middle years of the seventeenth century. Most are known to everyone who has any acquaintance with science at all—William Harvey, Robert Boyle, and Christopher Wren. The relatively unknown physician who pushed the studies forward most vigorously was the Cornishman Richard Lower.[1]

From the time of Aristotle and other Greek scholars, questions about blood in humans and animals have been addressed in various ways. Some have been metaphysical—attempting to work out, on the basis of a general idea about animals, the nature and role of blood. Some have been empirical—attempting to discover the nature of blood by experiment. In the seventeenth and eighteenth centuries the investigation of these topics was almost entirely experimental.

HARVEY'S EXPERIMENTAL DEMONSTRATION OF THE CIRCULATION OF THE BLOOD

The experimental programme of research into the nature of blood begins with William Harvey's demonstration that the blood circulates in a closed system of arteries and veins. Harvey (1578–1657) was Court Physician to Charles I, and is depicted in the famous painting by Holbein demonstrating the circulation of blood to the King.

William Harvey outlined his discoveries in a lecture series beginning in 1616. The material was published as *De Motu Cordis* in 1628. Recent scholarship has emphasized how far Harvey held on to aspects of the Aristotelian world view but it also shows how deeply he was committed to the experimental 'philosophy' of his own times. As an Aristotelian, Harvey framed his thinking in the pattern of the four 'causes'. Aristotle had set out a fourfold scheme that would need to be fulfilled if a complete explanation of anything could be achieved. The *aitia* ('causes' in the usual translation) included the material *aitia*, the stuff something was made of (blood); the formal *aitia*, the structure, form, or plan of the complete being (arterial/venous system); the efficient *aitia*, the active agent that brought it into being (heart); and the final *aitia*, what the being was for in some larger scheme of things. The final cause posed the question of blood's role in the life of an organism—what was the circulation of blood for? Harvey confessed he did know the answer.

His experimental programme began with a direct investigation of the relation between the contraction of the heart and the beat of the pulse in the blood vessels. By making a cut in the aorta he was able to observe directly that when the heart contracted, the blood spurted out of the incision under pressure. This was the phenomenon of the 'forceful systole'. The pulse was not an independent motion in the blood vessels, but an effect of the systole, the strong contraction of the heart. By making rough estimates of how much blood was ejected from the heart at each beat he realized that a certain more or less constant amount of blood was pumped around the body. Unless the blood circulated the total volume pumped out by the heart in a day would have been prodigious. So Galen's idea that arterial blood was absorbed in the bodily organs and venous blood created from the intake of food was clearly mistaken. There was just so much blood, moving 'outward' from the heart through the arterial system and returning via the veins. In the course of this process blood reached every part of the body.

Establishing the circulation of blood required the experimental demonstration of a number of major points. Not only did Harvey have to show how systole and diastole, detected in the pulse, were related to the contractions and relaxations of the chambers of the heart, but also to give a formal experimental proof of the way blood circulated to the lungs and back through the heart before being sent on to the rest of the body. In his lectures he emphasized that his claims had been established experimentally, that is, by dissection of living animals to reveal the circulatory system in action.

Summing up his researches Harvey says,

> It has been shown by reason and experiment that the blood by the beat of the ventricles flows through the lungs and the heart and is pumped to the whole body. There it passes through pores of the

> flesh [later shown to be minute capillary blood vessels] into the veins through which it returns from the periphery everywhere to the centre, from the smaller veins into the larger ones, finally coming to the vena cava and the right auricle.[2]

The shift from the use of indirect evidence and of arguments that such and such 'must be so' required experimentation on live animals. The incorporation of dogs in the apparatus became the accepted methodology of the physiologists of the time. Ingenious pieces conjoining the living bodies of dogs and other animals to metallic tubes, syringes, and non-living quills of feathers were constructed in the spate of experimentation that followed Harvey's experiments. The explorers of the next generation were intent of finding out the true nature of blood.

EXPERIMENTS ON BLOOD TRANSFUSION

The programme began with experiments on the effect of injecting various substances into the blood of a living animal. It soon moved onto a study of the possibility of blood transfusions.

At the core of the group of 'cognoscenti' who pushed through the experimental exploration of these possibilities stands the tall, slender, pale-faced, pious Robert Boyle, fourteenth child of the Earl of Cork. The centre of a network of correspondents, he collected, proposed, and criticized the experimental work of the other members, and carried through some of his own. For example, to test his atomic theory of matter as it could be applied to the vital processes of plant life he distilled the leaves of indoor plants he had grown entirely in water, or so he thought. The resultant oil was immiscible with water, showing that the corpuscles of water had been rearranged into a new configuration by chemical processes within the plant itself. Thus, he hoped to demonstrate a very general principle that encompassed non-living as well as living material. He concluded from this and a host of other experiments that the properties of material things arise from the arrangements of their insensible parts. Changing that arrangement led to a change in the observable qualities of things. The attributes of flesh and blood too were the result of the arrangement of minute corpuscles. The experimental method did not recognize the boundary between the living and the non-living, or between plants and animals.

During the several years that the experiments on blood transfusion continued Richard Lower reported the results of his experiments directly to Boyle. These can be found in a series of letters reprinted in Boyle's collected works; other experimenters also informed Boyle of the results of their work. In turn Boyle passed on these discoveries and techniques to a great many correspondents.

Robert Boyle was born into a wealthy Anglo-Irish family in 1627. His mother died when he was 3. However, despite his apparent delicacy he was sent off to boarding school at Eton at the age of 8. After four years he returned to the family estate to be educated by a series of tutors. In 1638 he was in Geneva studying not only mathematics but gaining proficiency in French and tennis. After a brief spell in Florence through various mischances he was obliged to spend two years in poverty back in Geneva. There he had a mystical experience that convinced him that he had been spared to use the sciences to demonstrate the existence and goodness of God.

Returning to England in 1645, inspired by this experience, he began the long running and highly productive scientific work that made his name, in spite of the Civil War. Making his headquarters in Oxford and retreating occasionally to his nearby estate he drew together a remarkable group of talented men, including

Christopher Wren and Richard Lower. In 1668 he moved to London, where he lived until his death on 23 December 1691. Boyle was a notable hypochondriac and worrier. He was constantly troubled by the possibility that he had inadvertently committed a sin of which he had been unaware, but for which he would be accountable at the Last Judgement. Concerned about his health in the changeable climate of central southern England he took elaborate precautions to avoid a chill. He possessed a wardrobe of cloaks of various weights and designs, each appropriate to a certain type of weather. He never moved without a selection of these garments.[3]

Robert Boyle, Christopher Wren, Richard Lower, and other members of the Oxford scientific circle, the self-styled 'cognoscenti', accepted Harvey's claims wholly. Taking the circulation of blood for granted enabled a number of questions about the nature of blood to be formulated more carefully than ever before. How is it aerated in the course of passage through lungs? How is its heat dissipated so that the body temperature remains more or less the same? How does it carry nutritive substances around the body?

In discussions that began around the years 1650–1 Boyle and his friends gradually formulated a workable research programme. The first step in trying to answer these deep questions would be to try to follow in detail how blood circulated.

In this vacuum of hypotheses the development of the research programme led to a series of experimental *explorations*, such as 'Could we nourish a dog by infusing broth directly into the blood system?', 'What would happen if we infused alcohol or milk into the dog's bloodstream?' However, to answer the myriad questions that the problem of the nature of blood threw up, there had to be something which could be handled and manipulated—there had to be a workable apparatus.

Since blood is a constituent of animals, there had to be at least one animal source in which blood could be studied by carrying out some plan of manipulations. Better still, why not take a living animal in which blood is already circulating and see what happens when various substances are introduced, by injection, into the 'mass of blood' of the animal? Would it be possible to create a hybrid apparatus, part non-organic (metal, etc.) and part dog, which would enable such experiments to be carried out?

The story of the development of a workable injection apparatus begins with Christopher Wren.[4] His investigation into ways in which various liquids could be *injected* into the bloodstream of a dog became the first developmental task of the cognoscenti—to refine the equipment so that liquids could be routinely and easily injected into an animal, and the animal survive. By following the effects of the different liquids on the dog a great deal might be learned about the nature of blood. Wren thought that the practice of injection might even have medicinal uses, as indeed it still has.

Christopher Wren was born on 20 October 1632 at East Knoyle in Wiltshire. His father, also Christopher Wren, was a well-known clergyman rising to serve as chaplain to Charles I. Young Christopher's mother died when he was only two. He was said to have been a delicate child. Despite the political troubles of the time, in which his father was caught up, he managed to acquire a good education. He was taught mathematics by his sisters and a local clergyman. At some time during his childhood he attended Westminster School in London. In 1645 he left school altogether to become an assistant to Sir Charles Scarborough, the medical anatomist. Wren's job involved the preparation of specimens for lectures. Here he acquired exceptional skill in manipulating biological material.

In 1649 he came up to Wadham College, Oxford. His choice of college was probably due to the fact that John Wallis, the foremost mathematician of the day, was Warden. The two began several collaborative projects, including designing various machines. After graduation he spent several years as a Fellow of All Souls College. During this time he was intimately involved with various cognoscenti projects. In 1657 he left Oxford for London to be Professor of Astronomy at Gresham College. By 1661 he was back in Oxford as Savilian Professor of Astronomy, a post he held for twelve years.

Wren's precocity was almost Mozartian. He changed professions with facility, displaying the same extraordinary degree of intellectual power and practical skills in whatever he undertook. During his time as Savilian Professor of Astronomy he designed several major architectural works that included original structural features such as the roof trusses in the Sheldonian Theatre at Oxford University. He also pursued anatomical and physiological interests with his fellow cognoscenti, published important work in mathematics, and even found the time to design agricultural machinery. The Great Fire of London of 1666 gave him the opportunity to engage in architecture on a grand scale—no less than the rebuilding of the city. Some of his plans matured; witness the great number of 'Wren' churches in the capital. He was appointed Surveyor General in 1674, continuing in that post until 1714. He died on 25 February 1723.

Before turning away from mathematics to architecture Wren busied himself in experimental researches in biology. His training as an anatomical assistant and remarkable skill in preparing specimens allowed him to find a way to incorporate dogs as an essential component in the apparatus necessary to study the circulation of blood and draw out its consequences. Among these experiments were those in which successful techniques of injection were developed, as the complex experimental apparatus was refined in various ways.

Summarizing Robert Boyle's account of the story, it seems that in the course of a discussion Wren claimed to be able to 'easily contrive a way to convey any liquid poison into the mass of blood'. In the event, with the help of various friends, he did manage to insert the thin pipe of a syringe into a ligatured vein, and injected a 'warm solution of *Opium* in sack [Spanish wine]'. Reading between the lines the phrase 'tortured dog' suggests that the friends' main role was to hold the dog still. The dog was stupefied for a while, but recovered and seems to have flourished. Unfortunately, as Boyle reports, the dog became famous and was stolen! The total apparatus then consists of syringe and dog connected by the pipe, very like a chemical experiment with flasks and glassware. Both components of this equipment were refined. The injection worked best with a large, lean dog with easily accessible veins. A bladder attached to a quill was more effective than a syringe. No doubt a quill could be cut so as to be very sharp. The rapidity with which the opium stupefied the dog demonstrated the overall correctness of Harvey's 'circulation' hypothesis. Wren's other experiments infusing wine at one time and an emetic at another had their characteristic effects with the same rapidity.[5]

Medical applications were talked about, including providing nourishment intravenously. The only experiment reported from this time involved injecting broth into the bloodstream. Richard Lower attempted this with no success. Milk and other fluids were used, but the project of intravenous nourishment was abandoned. Of course one vital fact about digestion had eluded the experimenters; while alcohol passes unchanged into the bloodstream from the stomach, the nourishing components of broth are the products of complex chemical processes, during which the constituents of the broth are for the most part broken down into simpler molecules. The increasingly sophisticated techniques of injection enabled Wren to experiment not only with wine but also with the injection of medicines such as opium directly into the blood.

However, more to the point scientifically, the injection of coloured liquids into the venous system just at the moment at which an animal was killed, allowed the whole of the blood vascular system to be observed in all its finer ramifications, except of course for the fine blood vessels, called capillaries. These had been observed in the lung of a frog by Marcello Malphigi (1628–94) some years earlier, and this work was known to Wren and Lower. An article by Malpighi describing his observations of capillaries in a frog's lung had been published in the *Proceedings of the Royal Society* in 1661.

If wine, opium, milk, and whey, why not blood? Injecting blood is no more or less than a 'transfusion'. Now the story shifts from an open-ended exploration with the perfected apparatus to a renewed attention to the technique by which transfusion or 'injection of blood' might be undertaken. In hindsight we can see that by not taking the concept of blood groups into consideration, a concept quite unknown to any of our cast of characters, most transfusion experiments were bound to fail, even if the apparatus was perfected.

Transfusion might be achieved in two rather different ways. The first, which has come to dominate modern transfusion techniques, involved taking blood from one animal or human being, storing it, and then injecting it into the recipient. The other technique involved linking the blood systems of two animals so that the blood from one, taken from an artery, would transfuse into the other, running into a vein. The necessity to take account of relative pressure of venous and arterial blood had been grasped only a year or two earlier.

The apparatus was developed by Richard Lower. He was born in Tremeer St Tudy on 1631 and baptized the following year on 2 January. After Westminster School he came up to Christchurch, Oxford, in 1648, staying there until 1662. Summing up the character of the man, Frank describes Lower as 'a brilliant and sometimes impatient technician. Adept at anatomy and vivisection, and proficient at criticizing specific hypotheses in physiology.' He brought his studies together in *Tractatus de Corde*, published in 1669. Lower had been much involved in the injection programme. His interest in transfusion seems to have been a natural step from injecting all sorts of liquids into animals, to trying the procedure with blood. And what better source of blood than another living animal? Lower pretty much developed the technique on his own.[6] Early attempts by Boyle and others had failed because they proceeded by taking blood from one dog into a bladder and then using a brass nozzle to inject some of it into another. The blood coagulated and blocked the instruments.

Lower's experimental programme worked out very well. In the year 1665 he succeeded in overcoming the blood coagulation problem by making sure that the source of blood was at a higher pressure than its destination. These experiments were interrupted by his efforts to find a wife with sufficient means to support him. Having been turned down in July by a widow in Garsington, a village near Oxford, he set off back to Cornwall in August. This time his suite seems to have been successful and he returned to Oxford with his spouse in February 1666. The troubles of the day delayed the spread of the news of his earlier successes. However, in 1666 he described them in a series of letters to Boyle.[7] The first improvement was the abandonment of any attempt at vein-to-vein transfer. The slowness of this method was partly responsible for the congealing of blood in the tubes. Instead all further experiments involved artery-to-vein transfer, since the pressure of arterial blood was greater than that of venous blood. The apparatus with which success was finally achieved consisted of three dogs, with quills to take blood from two donors and transfuse it into the recipient. A further refinement involved reducing the amount of blood in the recipient dog by bleeding it from the jugular vein before beginning the transfusion process. Then blood was taken from the cervical artery of one of the donor dogs and transfused to the jugular vein of the recipient. By repeating this procedure several times, Lower believed that he had replaced the blood in the recipient dog completely with blood from the two donors, who not unexpectedly had perished through blood loss. Lower describes the outcome in a definitely jaunty manner. Once the wound was sewn up the dog 'promptly jumped down from the table, and apparently oblivious of its hurts, soon began to fondle its master, and to roll on the grass to clean itself of blood... [showing little] discomfort or... displeasure'.[8]

Richard Lower continued to carry out extensive anatomical and physiological studies, including careful dissection and description of the muscles of the heart until his death on 17 January 1691.

BLOOD PRESSURE

By the end of the seventeenth century the transformation from the geometrical analysis of physical phenomena that one finds in Galileo's *Two New Sciences* of 1634 to their arithmetical treatment based on numerical measures had moved a very long way. Physicists and engineers were able to measure and to calculate important aspects of natural phenomena, such as length, volume, weight and mass, velocity, and pressure. The achievement of metrical precision, long a characteristic of astronomy, had begun to spread further into many scientific projects.

Reflecting on his engagement 'into this large field of statical and other experiments', Hales declared that

> whence we see great encouragement we have to spur us on in these pursuits; since the wonderful works of the great Author of nature are so fruitful in furnishing us, from its inexhaustible fund, with fresh matter for our researches, and thence with the inexpressible delight, of new and farther motives to adore and praise our all-glorious Maker in his works.[9]

Stephen Hales was born 17 September 1677 at Bekesbourne in Kent. He came from a family of distinguished public servants. His mother and father both died before he was sixteen, and he was cared for by his elder brother. In 1696 he entered Bene't College in Cambridge. The college had a particular interest in the biomedical sciences. Research was encouraged, among the undergraduates as well as the Fellows. Hales's proficiency in theological studies ensured his election to a Fellowship in 1703. In the manner of the time he was ordained priest in 1709. Effectively this guaranteed him both an income and a place in society for the rest of his life. Unlike many of his contemporaries he used the opportunity to carry on a long series of scientific research programmes.

His interest in biology was sparked by the arrival in 1703 at Bene't College of William Stukely. Stukely took advantage of every opportunity to study scientific topics experimentally. He and Hales quickly became friends. A contemporary describes them as 'rambling over Gogmagog Hills and the bogs of Cherry-Hunt Moor to get simples [botanical specimens] . . . They proceed also to the dissection of dogs. They applied themselves to chemistry.' Stukely left Cambridge in 1709 and Hales left soon after to take up the perpetual curacy on the parish of Teddington. He married Mary Newce in 1720 but she died within a year.

After her death not only did he continue all kinds of scientific work, but was one of the first to apply scientific methods to matters of public importance, such as the provision of clean air in ships and prisons and plans for the amelioration of the evil effects of the gin trade. He became a Trustee of the Colony of Georgia. Perhaps because of his interest in the conditions aboard ship, he set about devising ways to ensure that there was fresh air in the lower decks, developing a system of fans to force air into the confined spaces of the ship's interior. He insisted on the installation of ventilators in prisons as well. A Fellow of the Royal Society since 1722, he later served on its Council. In all he was a man who not only expressed an interest in the welfare of his fellow human beings, but vigorously pursued practical projects to achieve it. He died on 4 January 1761.

Stephen Hales endeavoured to bring the measurement of physical parameters to biological objects and processes. Just as Boyle had focused on the relation between the pressure and volume of gases, so Hales used the bodies of animals, manometers, and buckets for the same purpose in biology. What is the relation between the pressure and volume of blood and that of sap? In the first volume of *Statical Essays*, *Vegetable Statics*, Hales describes a number of experiments on the movement of sap in plants, in terms of rate of transpiration, pressure, and so on. In the second volume, *Haemastatics*, he describes his experiments with living horses to study the workings of the arterial and venous systems, particularly the question of blood pressure and rates of blood flow in different parts of the system. The aim was to achieve for biological phenomena what Newton and Boyle had achieved for physical phenomena. The metrication of chemistry had to await the new methodology of the French and English chemists during the last quarter of the eighteenth century with the development of the chemical balance.

In his dedication to the *Haemastatics* (vi–vii) Hales declares,

> As the beautiful fabric of this world was chiefly framed for and adapted to the use of man, so the greater insight we get into the nature and properties of things, so much the more beneficial will they be to us, the more will our real riches thereby increase, the more also will Man's original grant of dominion over the creatures be enlarged.

Within this framework of thought he carried out his research into the 'Hydraulics and Haemastatics' of the blood systems of animals. An ironic commentary appeared in Thomas Twining's poem *The Boat*, a pastoral piece set around the river Thames:

> Green Teddington's serene retreat
> For philosophic studies meet,
> Where the good Pastor Stephen Hales
> Weighed moisture in a pair of scales,
> To lingering death put Mares and Dogs
> And stripped the skin from living Frogs.
> Nature, he loved, her Works intent
> To search, or sometimes to torment.

Hales knew, as did most people, that the muscles increase in volume when active. How is this quantitative change to be explained? He tried to find out whether the expansion of muscles on contraction was due to inrush of blood pumping them up, or to some other mechanism, the expansion of the muscle material itself.

In the same volume, Hales describes in detail his research into the nature of the calciferous 'stones' or 'calculi' that when lodged in the bladder or the kidneys were a common and painful disorder at that time. A substantial prize was offered for a procedure that would dissolve these objects. The only alternatives were painful and dangerous extractions via the urethra or an abdominal operation of course without anaesthetic—'cutting for the stone'. Just as in his studies of blood pressure and the behaviour of muscles, Hales made use of animal models in his studies of this affliction, but most of his experiments were concerned with finding an agent to dissolve them that was not harmful to the patient. In this he was not successful.

The second volume of the *Statical Essays* is an example of the use of animal bodies as exemplars of the biology not only of brute beasts but of the human species as well. Lessons were drawn from these studies for the better performance of medical procedures of the time, such as blood letting. The Revd Stephen Hales did not slaughter hundreds of horses, but pursued his studies under the discipline of intensive design. Just a few old nags on their way to the knacker's yard were sacrificed in the interests of science. Hales' few old horses represented the whole of warm-blooded creation. In this chapter it will become clear that Hales' experiments should be read in the 'logical' framework as set out in the introduction to his second volume. He was ready to generalize from his discoveries with a few examples not only to all horses, but to the whole kingdom of *mammalia*, including human beings, though he did do a sort of cross-check of his intensive design by carrying out similar experiments on dogs and sheep.

William Harvey had shown that the blood circulates in a closed system of vessels. However, the details of the 'parameters' of the process were more or less unknown. The idea of circulation led to a clearer idea of the role of the heart. It must act as the pump which forces the circulation of the blood. That idea led very naturally to the question of the pressure of the liquid in the system round which the heart-pump drove the blood and the velocity with which it did so. Also, there was the connected question of how blood pressure was related to the volume of blood in the system. Was this relation analogous to Boyle's Law, in which these two quantities were inversely proportional to one another? What level of pressure did the heartbeat induce in the blood vessels? Was the pressure constant or did it vary with the phase in the cycle of the beating of the heart, between systole and diastole?

Stephen Hales' research programme was aimed at forming an accurate idea of pressure and other physical properties of the blood in a variety of animals. His first project was to make comparisons of blood pressure based on differences in the ratios of the size of the animal to its pulse rate. The rate of the heartbeat was more rapid in small animals, roughly proportional to the overall size of the creature. However, Hales found that the smaller the animal, the lower the blood pressure. It follows Bacon's Principle of Absence in Proximity that blood pressure is not a function of the rate of the pulsing heart. The only relevant physical variable, thought Hales, is the quantity of blood pumped out at each beat.

At Cambridge, with his pupil, William Stukely, he had filled the lungs of a dog with lead. After the flesh had rotted away a perfect replica of the trachea and alveoli was left. Using a similar method for the measurement of the volume of the heart, by filling the ventricle of a fresh but empty heart with enough wax to open the valves, he was able to make an estimate of the amount of blood expelled at each beat. The ventricles of smaller animals expelled proportionally more blood at each beat than did the hearts of larger animals.

So far so good—but was the blood pressure uniform throughout the system of arteries and veins? Intent on investigating the fine details of the blood systems in general, Hales made experimental comparisons between the rates of blood flow in different parts of the arterial and venous network. A rough calculation estimating the area of the alveoli in the lungs led him to predict that the rate of flow in the capillaries of the lungs would be much greater than in similar vessels in other parts of the animal body. This led to an experiment in which he compared the rate of blood flow in the lungs with rates elsewhere. Using a microscope and following the movement of single corpuscles in capillaries he showed that blood flowed about five times as fast in the lungs as elsewhere. Of course this experiment involved opening up the bodies of living animals to reveal sections of lung carefully enough dissected to allow the examination of single capillaries under the microscope.

The experiments for measuring blood pressure required the insertion of tubes into the arteries and veins of living horses. These were attached to simple manometers, pressure gauges, essentially long vertical tubes, to measure the pressure of the blood in the system. The horses, chosen from those on their way to slaughter, were tied to gates as suitable immobilizing frames. Here is Hale's account of one of his experiments:

> In December I caused a *mare* to be tied down alive on her back [to a field gate]; she was 14 hands high, and about 14 years of age, had a

> fistula on her withers, was neither very lean nor yet lusty: having laid open the left crural artery about 3 inches [76 mm] from her belly, I inserted into it a brass pipe whose bore was 1/6 of an inch [4 mm] in diameter; and to that, by means of another brass pipe which was... adapted to it, I fixed a glass tube of nearly the same diameter which was 9 feet [2.7 m] in length: then untying the ligature on the artery, the blood rose in the tube 8 feet 3 inches [2.51 m] perpendicular above the level of the left ventricle of the heart.[10]

Hales reports that at first the blood did not attain its full height, but increased gradually at each pulse. Once the full height had been reached it would rise and fall anywhere between 2 (50 mm) and 14 inches (355 mm) with each pulse. The normal pulse of a horse is about 36 beats per minute, but in this horse, notes Hales, 'being in pain' it rose to 55 beats a minute and sometimes to as much as 100.

So much for systolic and diastolic blood pressure, but how was the volume of blood related to the blood pressure? 'After each quart of blood was run out, I refixed the glass tube to the artery to see how much the force of the blood was abated.' He continued reducing the volume by quarts and then, after the eighth extraction, to pints. The pressure did not decline completely smoothly but by somewhat uneven steps. After 17 quarts (19 L) had been lost, the horse expired.

The table of results shows 'that the decrease of the force of the blood in the arteries, was roughly proportional to the several quantities of blood that were evacuated... This disproportionate inequality in the several heights was principally owing her violent straining to get loose', which made the blood in the fourteenth trial rise higher than it had done in several of the preceding ones (Figure 6). These spurts of higher pressure were also correlated with 'deep sighing', because, argued Hales, 'the blood passed

HÆMASTATICS. 11

The ſeveral trials.	The quantity of blood let out.		The ſeveral heights of the blood after thoſe evacuations.	
	Quarts	Pints	Feet	Inches
1	0	1	9	8
2	1		9	8
3	2		9	5½
4	3		8	4
5	4		8	2
6	5		7	8½
7	6		7	1
8	7		7	6½
9	8		7	4½
10	9		6	6½
11	10		6	7¼
12	11		5	11
13	12		* 5	8½
	12		† 4	5½
14	13		4	4
15	14		3	8
16	14	1	‡ 4	2
	14	1	§ 3	2
17	15		3	3½
18	15	1	2	10

* The higheſt point it would ſtand at for ſome time.
† The loweſt points, at which points it would continue for ſome time.
‡ The higheſt point.
§ The loweſt point.

He did not bleed half a pint more after this laſt trial, before he expired.

Figure 6. The relation between volume and pressure of the blood in a horse.

from the more actively moving lungs in greater quantity into the left auricle and so to the ventricle from which it was pumped at a momentarily greater pressure.' Hales conjectured that this is why sighing, by increasing the force of the blood, cheers up those who are dejected and sad.

By draining the remainder of the blood from the horse and estimating losses during the insertion of the tubes which acted as a pressure gauge, Hales estimated that the quantity of blood in a horse is at least 1,105 cubic inches (18 L).

Hales draws various lessons for managing the medical practice of 'bleeding', for example how much blood can safely be taken. If medical theory suggests a great quantity, then these studies show that it should not be taken all at once.

He also carried out similar studies on sheep and dogs. Of course the procedure in each case culminated in the 'expiry' of the creature being used.

Thomas Twining's equivocal poem refers to Hales 'stripping the skin from living frogs'. The idea of thinking about 'blood pressure' in terms of the effects of pressure in other physical phenomena had led to other questions. For example, was the swelling and contraction of muscles due to this pressure, as a bladder swells when air is pumped in? Only when the muscles were laid bare of any covering could the contractions and relaxations of muscle fibres be directly observed. His observations led him to reject the theory that it was blood under pressure that inflated the muscles. The blood supply to the muscles ran along the grooves between the muscle fibres rather than feeding into the fibres themselves.

> If the skin be flead [*sic*] off the belly of a live frog, and the abdomen opened on each side, so that its strait muscles may, by drawing a little to one side, have a strong focal light cast on the inside of them; if in this posture those muscles be viewed thro' a good microscope, the parallel fibres of the muscles are plain to be seen, with the blood running up and down, between each fibre, in capillaries so fine only a single globule can pass them. If the

> muscle happens to act while thus viewed [for example by touching the foot of the frog], then the scene is instantly changed from parallel fibres, to a short series of *rhomboidal pinnulae*, which immediately disappear as soon as the muscle ceases to act.[11]

Commenting on this experiment Hales remarks that it is likely that the muscle contraction is caused by action of the 'animal spirits' via the nerve fibres. He also observes that there is no sign of blood vessels running into the muscles. The hypothesis that muscle expansion is due to increase of blood volume was clearly mistaken.

From the point of view of the philosophy of experimentation, this is not a simple exploration, like the studies of the previously unknown relation, if any, between the volume of blood and blood pressure. It is also an 'either/or' test. However, as in all such tests, the alternative left standing is still not proven, and this example shows it very clearly. Showing that it is not the blood that 'pumps up' muscle volume does not entail that it must be influences from the animal spirits. Many other possibilities remain open, some sensible, others less so.

THE 'BLOOD OF PLANTS': THE HYDRAULICS AND STATICS OF SAP

Hales's construction of pieces of apparatus that were hybrids of organic and inorganic parts reached a high state of scientific elegance in his experiments on the rate of flow and pressure of sap in plants. Analogies between animal blood and vegetable sap had been commonplace for centuries. Hales undertook to test this idea by experiments and particularly by measurement. The description of these hybrid studies fills the first volume of the *Statical Essays*.

These are some of the most powerful and influential experiments ever undertaken in biology gauged in terms of their long-term influence, which compares to that of Mendel's experiments on heredity. Hales showed that the methods, techniques, and *concepts* of the physical sciences had a vital role to play in biology.

Nehemiah Grew (1641–1712) had revealed the existence of a network of vessels within the plant capable of transporting liquids and/or gases. Marcello Malphigi (1628–94) realized that plants like animals must breathe and that the atmosphere somehow contributed food to the plant. Clearly sap was in some ways analogous to blood, but how did it move in the plant—did it circulate or did it pass through the skin into the roots and out from the leaves? There was no sign of any sort of pump that might correspond to the heart.

The focus on numerical measurements is evident in the second experiment of the *Vegetable Statics*. Collecting the water that perspired off a cabbage plant Hales found that it measured 32 cubic inches (524 mL). The surface of the leaves was 2,736 square inches (0.9 m^2), so an amount of water of 1/86 inch (0.3 mm) in depth perspires in a 12-hour day. By calculating the cross-sectional area of the stem Hales declared that the ratio of the velocity of sap in the stem to the velocity of perspiring sap at the surface of the leaves is 2,736:4,268. However, the water of which the sap is largely composed must come in through the roots. By measuring the length of all the cabbage plant's roots and calculating their surface area he found that 'the sap must go with near eleven times the velocity through the surface of the roots, than it does through surface of the leaves.'[12] Since many roots are lost in transplanting, Hales advises that relocated plants should be very well watered.

To test the hypothesis that sap moved from roots to leaves powered by the transpiration of water, Hales took pairs of branches from a variety of trees, and stripped the leaves from one of each pair, placing them in separate glasses with the same measured quantities of water. The branches with leaves imbibed between 15 and 30 ounces (425–850 gm) in a 12-hour day, 'more or less in proportion to the quantity of leaves they had... while those without leaves imbibed but one ounce, and were heavier in the evening than the morning, they having perspired little.'[13]

In another set of experiments Hales set out to try to discover the mechanism by which the transport of water occurred. Was it driven by pressure from the roots or drawn up by transpiration at the leaves? To understand this experiment it is helpful to consult Figure 7, which shows his experimental setup.

Hales fixed a 7-foot (2.13 m) tube to the end of an apple branch with plenty of leaves and filled it with water. The experiment was to compare how far the water fell under the tube in each of two conditions. In each case the water in the tube was exerting the same pressure. When the leaves were immersed in water and could not perspire, the water fell very little as in Figure 7a. But when they were in the open air the water in the tube fell rapidly, as in Figure 7b. Since the pressure exerted on the sap in the branch was the same in each case the driving force of water transport must have been the transpiration from the leaves.

Here we have the apparatus by which various aspects of plant physiology are measured. In each case the device is a hybrid, part organic and part inorganic. At the same time this very apparatus is being used to measure the physical properties of the process of which it is an essential component.

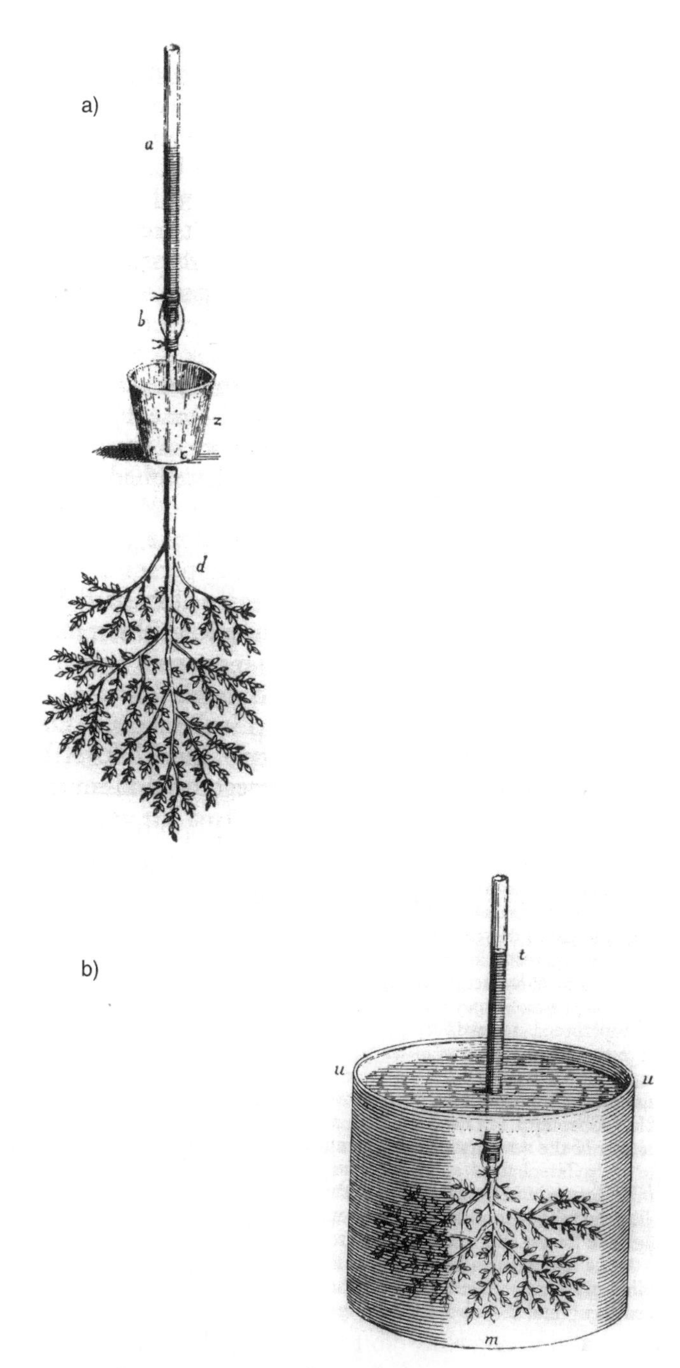

Figure 7. Leaf transpiration in and out of water.

Once again in both series of experiments the method is clearly that of the intensive design. Only a few examples are subjected to experimental study, and their typicality is the basis on which general and even universal claims can be confidently made.

SPHYGNOMETRY

It is evident that any attempt to use Hales's method for determining blood pressure in human patients would have been met with resistance, if not outright rejection. Whether Hales could have saved the horses which served as his laboratory equipment for his experiments is hard to say. In fact he made no attempt to do so. Here we have an invasive procedure par excellence. During the next couple of centuries various attempts were made to develop non-invasive techniques for these measurements. They depended on the fact that blood vessels could be compressed by applying external pressure so that the pressure in the vessels themselves would be revealed by resistance to this compression. The methods were complex and quite uncertain in their outcomes. The problem was solved by Nikolai Korotkoff.[14] In 1905, he published his method, essentially the one we use today. Though this is more applied science than original research the method he evolved is a direct descendant of the studies made by his English predecessors.

It is curious that the man who developed a procedure that is used all over the world is virtually unknown to the millions of patients whose blood pressure is taken routinely every day. Medical students learn about 'Korotkoff sounds' but I have never heard this phrase pass the lips of any of the doctors at my local clinic.

Nikolai Korotkoff (1874–1921?) worked in Moscow University in the Department of Surgery. The Russian Empire was expanding to the East and young Korotkoff twice found himself working as a physician in China. The first occasion was the Boxer Rebellion in 1900. His second visit to the East was during the Russian–Japanese War (1904–5), when he was sent to Harbin, in northeast China.

Driven by the needs of the wounded, Korotkoff was working to solve a problem already formulated in 1832 by Nicolai I. Pirogov. When a major arterial blood vessel damaged by injury is closed with a ligature how can we be sure that collateral vessels will be adequate to take over the circulatory role performed by the original vessel? While attempting to resolve this problem, he systematically listened to the collateral arteries to estimate their potential strength after a major vessel of the wounded limb had been closed with a ligature. He noticed that certain specific sounds could be heard during the decompression of the arteries. This auditory phenomenon, nowadays generally known as 'Korotkoff sounds', became the basis of the new method of blood pressure measurement.

To measure blood pressure, Korotkoff fitted an elastic cuff to the middle third of the patient's arm. The pressure in the cuff was increased until the blood supply to the periphery was completely stopped. Then the pressure in the cuff was decreased, and a stethoscope was used to listen to the artery 'directly below the cuff'. Once the pressure fell below a certain level, the first short tones could be heard. This indicated the passage of the first pulse wave along the artery below the cuff. The manometer or pressure gauge reading at which the first tone appeared corresponded to the systolic pressure, the pressure of the positive pumping stroke of the heart. With a further decrease in pressure in the cuff, the tones were replaced by murmurs that were followed in turn by second tones. Finally, all the sounds disappeared. The moment all sounds subsided, according to Korotkoff, blood was flowing freely through the arteries. The diastolic pressure in the artery at that moment slightly exceeded the pressure in the cuff, and the manometer reading at the moment the sounds disappeared corresponded to the diastolic pressure, at the resting phase of the heart's pumping cycle. Korotkoff's method of measurement of blood pressure quickly became a standard medical procedure. It plays an important role in the study of the etiology, pathogenesis, and treatment of hypertension.

Breath: the composition of the air

The fact that the air is not a simple substance was realized very early on in the modern era of experimental science. A candle would soon go out if left burning in a closed vessel, for instance a glass jar. The amount that the air had diminished by that time was easily measured by setting the jar open end down in water and setting the candle on a stand inside it. Atmospheric pressure drove the water level to compensate for the loss of pressure caused by the candle. Was this because a constituent of the air had been removed or was it because the natural elasticity of the air had been diminished?

In the seventeenth century the role of air in the sustaining of living beings was a matter of increasing interest. The first person, it seems, to explore this question systematically was John Mayow (1643–79), who perhaps has some claim to have isolated the substance we call oxygen and to have understood its role in combustion and breathing. Mayow realized that air consisted of two different kinds of particles, only one of which sustained combustion and life.

His experiments with nitre (sodium nitrate) had convinced him that 'the igneo-aërial particles common to nitre and air are not the air itself but very subtle particles which fixed in [that is, combined with] air and in nitre constitute their more active and fiery part.' There is an 'igneo-aërial spirit . . . fixed in the saline particles of nitre very much in the same way as in the aerial particles'.[15] In a way this 'spirit' was more or less what we think of as oxygen. However, Mayow clearly took the air to be a compound rather than a mixture, and so too was nitre.

Why did water ascend 'in a glass in which a lamp or an animal is enclosed although the air exists in it in the same abundance as before?' (84). His explanation was echoed a century later by Stephen Hales. Mayow argues that 'no other conception is possible than that the elastic force of the air has been diminished, and that this is due to a certain change wrought in the aerial particles themselves' (84). He goes on to reject the hypothesis that in vessels in which an animal or a lamp has been enclosed 'a part has been entirely consumed while the rest has remained unchanged, because if it were so there would be nothing to hinder the animal from still breathing in it' (84). His experimental apparatus consisted of a glass vessel in which an animal had been breathing until it expired. On opening the vessel to the atmosphere again he found that the formerly enclosed air still resisted the force of the atmosphere. Considering the mixture hypothesis again he remarks that if 'the elastic force of the air resulted from certain agile particles interspersed in it, and if these were exhausted by the breathing of the animal or the burning of the lamp then air in which an animal or lamp had been enclosed would by no means expand with as much force as unaltered air' (85); that is, both the initial unexpired and the subsequent expired air are at the same pressure. To test whether there was a difference between the expired air at the top and at the bottom of the vessel Mayow used two birds and later two mice—one of each pair was at the bottom of the vessel and the other at the top. The animal put in the upper part will 'for some time survive the other'. Mayow also notes 'that when a small animal, such as a mouse, is shut up in a glass and suffers from want of air, it turns its mouth hither and thither in an upward direction in quest of breath' (87).

The expired air rises, so Mayow believed, and in the upper atmosphere is refreshed with the proper compliment of nitro-aërial particles. This wise arrangement by the 'highest and best Artificer' is essential for the benefit of human society:

> the life of each would be a sort of perpetual pilgrimage, inasmuch as we should find it necessary to wander by night and by day, through the world and desert places, not so much to gain wealth and foreign dainties as hunt after aërial nourishment, and to banish ourselves far to avoid the popular breath. (88)

This passage hardly needs any comment in our time.

Nearly a century later Stephen Hales extended his studies to experiments designed to measure the amount that air changed by the effect of 'the breath of Animals enclosed in glasses'.[16] His experimental programme is described in an additional section added to the second edition of volume 1 of his *Statical Essays*. He states:

> I placed on a pedestal, under an inverted glass . . . a full grown Rat. At first the water subsided a little, which was occasioned by the rarefaction of the air caused by the heat of the animal's body. But after a few minutes the water began to rise, and continued rising as long as the Rat lived, which was about 14 hours . . . about 1/27 part of the whole [was absorbed]. (234)

In later experiments he found that a small rat used 2/13 of the air, while a cat used 1/30.

In this description Hales seems to be following John Mayow's thought that breathing changes the elasticity of the air rather than absorbing part while leaving another part behind. To find out what was happening to the air to account for the loss of elasticity, that is, the power to hold the water at bay in the jar, he experimented on himself. Securing a bladder to a tube, he blew it up and then 'pinched my nostrils close so that no air might pass that way, so that I could only breathe to and from the air contained in the bladder'. After a minute he found himself suffocating, and stopped the experiment. However, he noted that the bladder had become flaccid and his 'lungs were much fallen'. The air was losing its elasticity. 'A considerable part of the elasticity of the air is destroyed' in the vesicles of the lungs, he concludes.

In keeping with eighteenth-century thought Hales offers an explanation of the phenomena he has observed in terms of forces of attraction and repulsion. In the lungs the elastic particles of air, 'now changed from an elastsick [*sic*] repulsive to a strongly attracting state, may easily be attracted through the thin partition of the vesicles [in the lungs], by the sulphurous particles which abound in the blood' (248).

Commentary

The parallel between the studies carried out by physicists and chemists such as Robert Boyle and those of Hales and Mayow on the behaviour of fluids in living animals was clear to the scientists of the day. For the clergymen scientists of the time both were the work of God, providential, if we could only see it, for his creatures, particularly ourselves. The methodology was inductive, but not in the sense of a summing up of a great many instances to establish a law. The intensive design dominated the way evidence was sought and used. Results established on a few examples, taken provisionally to be typical, were good for all the members of the same species. Hales experimented on four horses and few other animals and only one trial was enough to show him that the leaves were the key to transpiration, the force that drew sap up from the roots to the top of the plant.

Chapter 5
Exploring a New Domain: A Bald Mouse, Two Chimpanzees, and Some People

Sometimes explorations are undertaken into domains of nature about which almost nothing is known. In this chapter we will look first at the apparently sci-fi idea of creating artificial organs to replace any that have been damaged or diseased. This project was taken up by the Vacanti brothers in their attempt to create the cartilaginous structure of the outer part of an ear.

The second of the exploratory experimental studies to be reported here was an attempt to bring up a chimpanzee capable of using meaningful signs in much the same way as human beings use words. Were the project to succeed it would answer the long-standing question of whether only human beings are capable of acquiring and using language.

The third experimental programme we will follow in this chapter was an attempt to answer the question of whether only the most depraved and evil people would engage in systematic cruelty to their fellow humans in response to the orders of officially appointed authorities. This thesis was tested by setting up a research programme in which ordinary people were ordered to give seemingly lethal electrical shocks to another human being, allegedly in the interests of scientific research.

These research programmes touch on deep and significant questions. It cannot be said that they have delivered answers that satisfied everybody. Flaws have shown up in the chimpanzee work, and serious methodological objections have been made to studies that seemed to show that depths of unthinking cruelty were easily accessed in very ordinary people. The work of the Vacantis was much misunderstood, though even the misunderstandings raised legitimate questions about the boundaries of scientific research, though not directly with the growing of replacement organs. Nevertheless, the results and the methods of research used in these studies are well worth pondering. In each case an apparatus was created in which a living being formed an essential part.

The creation of an artificial organ: the Vacanti mouse

The creation of the cartilaginous structure of a complex organ need only be accomplished once to prove for all time that this feat is possible. This was the role of the experimental work that led to the creation of the cartilaginous structure of an ear, nurtured on the back of a mouse. Here again the logic is that of the intensive design.

The line between engineering and science is not always easy to draw. In engineering people use their knowledge of the laws of nature to bring some practical project to fruition. For example, the science of mechanics is put to use by the designers of bridges, office buildings, and many other structures. The idea of 'tissue engineering' has been about for many years. Surely it should be possible to put biological knowledge to work to create new body parts that could be used to replace and repair damaged organs. Such a technique would have many advantages over the current procedures involving transplanting organs from one person to another. In the ideal case the artificial organs would be grown from body cells of the very person whose diseased or damaged organs need replacing. If this could be done there would be an indefinitely enlargeable reservoir of organs for transplantation. At the same time growing these organs from the cells of the person for whom they were intended would ensure that there would be no problem of tissue rejection, and hence no need to suppress the defensive activity of the recipient's immune system.

The most startling and, as it happened, notorious example of tissue engineering was the 'Vacanti mouse', a creature that seemed to have a human ear growing out of its back (Figure 8).[1] The living mouse was the apparatus with which the process of growing a substantial cartilaginous structure was sustained.

The ability to grow sheets of human cells in the laboratory had been developed many years ago. These hybrid materials—both natural in origin and artificial in production—have valuable medical uses in treating burns and other skin tissue damage. They are produced in apparatus that is almost entirely inorganic, glass and plastic. The organs of a human body are solid objects with elaborate internal structures. How could the techniques of cell growth in two dimensions be adapted to creating structures in three?

Joseph Vacanti, a specialist in surgical transplantation of human body parts, had long been troubled by the gap between the number of people needing replacement organs and the reservoir of available substitutes. He realized that if suitable scaffolding could be built, by seeding it with appropriate cells, a

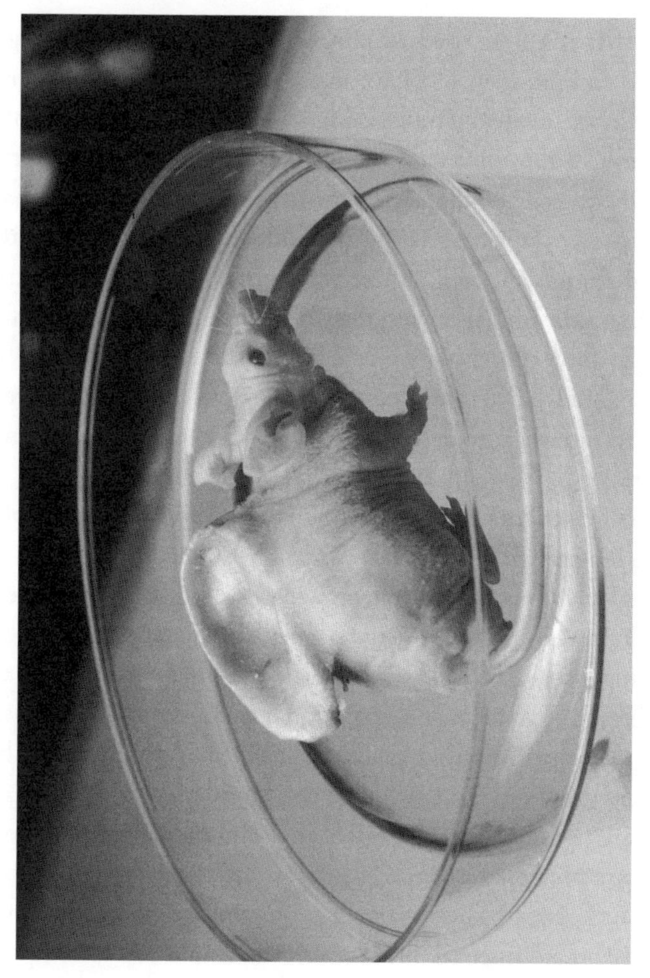

Figure 8. The mouse carrying the cartilaginous components of a synthetic human pinna.

three-dimensional, realistic surrogate organ could be created by the same technique as that by which sheets of cells were bred. The scaffolding had to be biologically inert, robust enough to serve as a template for the developing cell population, and in the end biodegradable. This process should leave only the cells appropriate for the artificial organ that could be substituted surgically for the diseased or damaged part of a human body. The story goes that Joseph Vacanti got the idea for building the scaffolding in a branching pattern from watching seaweed in shallow water. Using this structure as a template would enable cells to be seeded in a three-dimensional array from the beginning.

Joseph Vacanti is the eldest of four brothers, all of whom studied medicine and eventually became involved in tissue engineering. Their father, Charles Vacanti, was a dental surgeon and professor of dentistry at Creighton University at Omaha. The idea of growing a three-dimensional organ had occurred to Joseph Vacanti, but its realization involved his younger brother Charles, who had successfully grown human cartilage tissue in a three-dimensional structure. The key to the technique was the use of a scaffolding of biopolymers on which to seed the cartilage cells.

The third major contributor to the project was Robert Langer. He is well known for his work on the development of systems for the delivery of drugs and other medicines into the body in a non-invasive way. However, he has also been much involved in the kind of tissue cultivation and development that made him an invaluable colleague of the Vacanti brothers. From a purely chemical point of view the scientific aspect of the experiment turned on the ability to manipulate biopolymers. These are substances with a molecular layout that creates a ribbon-like form. They are the main constituents in the thread used for surgical stitching. Since they are organic though biologically inert, they gradually dissolve in the body. Langer had already begun to develop structures made from biopolymers, and as a close personal friend of Joseph Vacanti, took on the job of finding the materials for the necessary scaffolding.

A substantial mesh could be made from biopolymer fibres. After some trial and error Charles Vacanti and his colleagues were able to form the mesh into the exact shape of the cartilaginous framework of a human child's ear. The idea of growing the pinna of a human ear, a highly convoluted object, suggested itself as a demonstration of what could be done with the technique. Cartilage cells from the knee of an ox were used to seed the polymer mesh. The growing cartilaginous structure was supported physically by the biodegradable mesh to which the originating cells had been attached.

But why go to the trouble of transplanting the mesh and its seeded cells on to the back of a mouse? This step was necessary to ensure that the cells in the mesh should receive appropriate nourishment to enable them to divide and grow. By transplanting the structure on to the back of a hairless or 'nude' mouse a site was created in which blood vessels would grow from the mouse's body to provide nourishment for the 'ear'.

This particular mouse had not developed an immune system, thanks to a genetic defect. It was an ideal place to grow the surrogate ear, since without an immune system the nude mouse's body could not develop a hostile response to alien cells, in particular to the ox cartilage cells of which the artificial ear was formed. This structure 'took' very well, and a network of blood vessels developed along with the proliferating cartilage cells getting just the right kind of nourishment from the mouse to keep them replicating. In fact what was grown was the pinna, the outer part of the ear. In a sense this object was not an ear at all, since it did not include any of the auditory structures on which the ability to hear depends. After some months the biodegradable surgical stitch material of which the scaffolding was made dissolved away, leaving a cartilaginous structure able to support itself.

When this work was published in 1997 it provoked a variety of strong reactions. It was quickly realized that the work had opened the way for surgical techniques that had the potential to transform some people's lives. But thanks to a serious scientific misunderstanding there was also a chorus of protest. The 'Turning Point Project', a group hostile to genetic engineering, described the experiment as genetically engineering a mouse to grow a human ear on its back. Whatever may have been the rights and wrongs of the experiment morally, it had nothing whatsoever to do with genetic engineering. The DNA of the mouse played no part in the origin of the artificial pinna. The ear was built out of surgical stitching material and cells from the knees of oxen by the manipulative skill of the experimenters. The mouse contributed only a safe haven and a source of nourishment so that the ear might grow. The DNA of this mouse was special only in that it had been defective in the genetic basis of a functioning immune system.

The mouse was an apparatus, but one which, so to speak, 'enclosed' the process by which the ear grew, namely by the multiplication of the cartilage cells. It is always helpful to look around for an example of an apparatus built entirely of inorganic materials that performs the same function. Examples are not hard to find, even if the process to be supported is organic. The humble Petri dish, the shallow bowl in which bacterial cultures are grown, provides, like the mouse, a stable physical support for the process of cell multiplication. Most of the apparatus used by chemists, at least in the 'good old days', consisted of neutral, inert materials, glass being the most important. A chemical process could 'run' in isolation from any extraneous and potentially contaminating influences, whether the reaction was organic or inorganic.

'Talking' chimps

About forty years ago the idea of embedding a young chimpanzee in a human environment was proposed by Allan and Beatrice Gardner as a key phase in the investigation of the nature of language. The plan was to develop an experiment that would serve as a clear-cut test of the viability of an important hypothesis in linguistics. The experiment began in 1966 with an infant chimpanzee. They called her 'Washoe', after the river that runs through the city of Reno, Nevada, where the programme was sited. She was brought up in a human environment, living partly in the house with the Gardners, and having her own quarters in a 'mobile home' in the garden. A large dead tree nearby provided a kind of jungle gym.

The project was aimed at an exploration of the idea that a chimpanzee could, in suitable circumstances, acquire an understanding of an appropriate form of language that was of the same general character as human language. The Gardners, noting that chimpanzees in the wild use gestures for various purposes, built their research around ASL, the American Sign Language. In the course of this work they tested a significant principle enunciated by Noam Chomsky in his 1966 book, *Cartesian Linguistics*, and expounded in more technical detail in other publications. Chomsky claimed that language was species specific; that is, it was an attribute of human beings alone. His reasons for this claim were roughly that the capacity to extract language from an acoustic environment was innate. Each person inherited a version of the LAD (Language Acquisition Device) which 'searched' the auditory environment for language-like forms. Chomsky's account of the foundations of language involved both 'competence', that is, a body of linguistic knowledge, and 'performance', putting this knowledge to use in speech and writing. His well-known 'transformational grammar' was, in the first instance, a set of rules by which any grammatically well-formed sentence could be generated from a common base structure. How this sentence might be used was another matter.

Language is a system of signs, which are *shaped* and *used* according to a limited repertoire of rules of correctness. Chomsky's 'rules of transformational grammar' express the repertoire of syntactically correct sentence forms. These are the rules for the sequential organization of individual signs into sentential structures, according to the word classes presumed by language users. For example, most languages juxtapose a noun or noun phrase with a verb or verb phrase. Two successive nouns do not, in general, stand up as a well-formed sentence.

There are also rules for using signs meaningfully. Some govern descriptive uses of language, that is, for giving true or false accounts of how things are, or seem to be. Some express local conventions for performing all sorts of social acts, ranging from the intimate such as apologizing for a mistake, thanking someone for a favour, to the massively influential such as a Presidential declaration of war on another nation. These are Wittgenstein's 'rules of use' as described in detail in his *Philosophical Investigations* of 1953.

The use of the word 'rule' for expressing the norms of correct language use, be it speech or writing, is linked to the idea that scholars can describe the norms for a language in terms of two sets of rules, the dictionary definitions of words and the grammar for their employment in sentences. Roughly the first group covers semantic and performative uses, that is, the use of words in descriptions and their use in performing social actions, while the second covers the formation of acceptable syntactic structures.

To have a language an individual must manage signs competently according to both kinds of rules. They are explicit expressions of the implicit norms of the local language culture. The claim that only human beings were capable of this kind of management of sign systems was the hypothesis that the Gardners' explorations of chimpanzee capabilities was indirectly putting to the test.

The Gardners' programme of research was based on a close study of a single animal, a young chimpanzee. They believed that this project, if it were to be successful in showing that Washoe used signs in such a way as to distinguish the meanings of sign clusters in which the same signs were produced in different sequences, would not only reveal the scope of chimpanzee capabilities but also refute Chomsky's principle of species specificity. The logic is very simple. If someone makes a strong claim that *all* A's are non-B's, a single A that is a B refutes the claim absolutely. Chomsky's claim was forceful. According to him all animals, except human beings, are non-language users, in the sense that they are incapable of acquiring the specific skills that we know as language. One chimpanzee that does acquire these skills, in a reasonable measure, refutes the claim. Though logic speaks sternly, declaring that the truth of the probing statement entails the falsity of the hypothesis probed, in most cases the upshot is a flurry of qualifications, modifications, and restrictions of the original hypothesis, as well as a scrutiny of the methods by which the contrary evidence had been found. Did the Gardners refute Chomsky's claim? As we shall see, 'reality' turned out not to so clear-cut. How much is a 'reasonable measure'?

The way the sign language was learned was meant to parallel the way people learned a verbal language—partly by rewarding spontaneous gestures that are somewhat like the signs of ASL, sometimes by imitation, and partly by explicit instruction in how to use and to shape the hand into a meaningful sign. Washoe, the infant chimp, produced gestures like the signings of ASL. These were 'rewarded' in relation to the sense the sign would have had, had it been genuine. Washoe also imitated the signs used by the Gardner's assistants, and used them to get what she wanted. All this was supplemented by deliberate shaping of the infant's hand into a formal sign which brought forth its appropriate response.

This was unknown territory. The programme, the apparatus for which was an animal, whatever its logic, was an attempt to find out what had hitherto been unknown, the linguistic capacity of non-human primates from a single case study—the intensive design 'in spades' one might say!

Over a period of six to seven years Washoe acquired a 'vocabulary' which, variously measured, numbered more than 200 distinct signs. What is more she spontaneously developed new signs by a kind of metonymy. Drawing the right hand back and down with the left hand clasped expressed the act of peeling a banana. The Gardners were delighted to see Washoe using the same sign for turning on the tap for the garden hose, and for opening the back door of the station wagon preliminary to going for a ride. She generalized the meanings of other signs as well.

The way Washoe acquired the sign for 'more' is of special interest because it was one of the signs that Herbert Terrace trained his specimen chimp to use by a similar method. Chimpanzees like being tickled. The Gardners seized on this as an easily offered reward.

> We noted that, when being tickled, [Washoe] tended to bring her arms together to cover the place being tickled. The result was a very crude approximation of the ASL sign for 'more' . . . we would

> stop tickling, and then pull Washoe's arms away from her body. When we released her arms and threatened to resume tickling, she tended to bring her hands together again. If she brought back together we would tickle her again. From time to time we would stop tickling and wait for her to put her hands together by herself... Soon a very good version of the 'more' sign could be obtained, but it was quite specific to the tickling situation.[2]

Later, with other pleasing games as the focus the Gardners were able to extend the use of the 'more' sign 'to all activities including feeding'. Eventually Washoe would use the sign whenever there was a pause in a pleasurable activity.

Clearly an experiment was afoot. But what exactly was the apparatus? It extended far beyond the individual chimp to include the everyday environment in which the sign language was a potent instrument for getting things done. Paradoxically it both included and did not include the Gardners. In so far as they were model signers providing the conditions for the acquiring of ASL they were in the experiment and part of the 'apparatus'. In so far as they were observing and filming these activities they were not in the experiment and not part of the apparatus. They were playing the same role as the scientists peering into the test tube where the reaction was slowly taking place, or looking down the microscope to follow the life course of a bacterium.

And yet, despite heroic efforts to find at least a few cases in which the order of the signs affected their meanings, Washoe's capacity to use multiple signs in consistently orderly groups did not seem to develop in the way that such order develops with the growing language capacity of a human child. In short Washoe's semantic aptitudes, her capacity to express intentions, wants, and needs, were the limit of her linguistic development. There was some kind of semantics, but a marked absence of syntax. In a 1969 paper the Gardners offer a cautious conclusion apropos of the Chomskian thesis of species specificity: 'Theories of language that depend on the identification of aspects of language that are exclusively human must remain tentative until a considerably larger body of intensive research with other species [and indeed with chimpanzees] becomes available.'[3]

The Gardners themselves believed that they had found evidence of syntactical order in at least some of Washoe's performance of clusters of signs, two- and three-unit groupings. This has turned out to be controversial. The unravelling of the problem of primate 'syntax' is due to Herbert Terrace, a psychologist at Columbia University.

During the 1970s Herbert Terrace set about a programme of experimental work with the ironically named Nim Chimpsky, like Washoe an infant chimpanzee. Terrace's project was sharply focused on the question of the production of sentences. It involved very careful and painstaking statistical studies of the patterning of two-member sign clusters in relation to the contexts in which Nim produced them. Did Nim distinguish between the meanings of 'hat me' and 'me hat'? In short did syntax play a part in Nim's grasp of a rudimentary language? Terrace and his assistants taught Nim signs in much the same way as the Gardners and their assistants had taught Washoe. The apparatus was made up of the same living components, including the trainers, together with the 'inorganic' recording equipment. There was Herbert Terrace, his crew of assistants, and the infant chimpanzee Nim. Looking on from afar so to speak was a second incarnation of Terrace, the experimenter manipulating the equipment and studying the results.

Terrace summed up the outcome of the statistical analyses of Nim's signing abilities as follows: it seemed that 'a solid basis for demonstrating that a chimpanzee can create sentences [had been established].' The statistics seemed to show that Nim used sign order meaningfully; that is, he used 'hat-me' and 'me-hat' differently—but was this difference meaning oriented, that is, for example between a request and an assertion—'Give me a hat' and 'That hat is mine'? 'The data invalidate two simpler interpretations of word [sign] sequences emitted by a chimpanzee; that they were learned by rote or that they were random combinations of signs, each relevant to a particular context but unrelated to one another.'[4] But a happy result turned out to be erroneous. Something else was going on.

What underlay the illusion of sentence formation was revealed only by a very close study of the moment-by-moment sequence of signings by the teacher and Nim, the trainee. Terrace's suspicions had been aroused by such unchild-like features as the very small number of signs Nim used with the same meaning. For example, he reports that '90% of the beneficiaries in utterances judged to be object–beneficiary combinations were Nim and me.' When the video tapes were studied closely, and attention was focused on both the teacher's and Nim's signs, it was evident that Nim was overwhelmingly often repeating the sign that the teacher had already made—sometimes simply, and sometimes with a certain amount of elaboration. There was little spontaneity of utterances, indeed only 10%. Furthermore, though Nim's signs reflected the prior signings of the teachers, the teachers' signs did not seem to be semantically very significant, in that Nim interrupted the teachers' signings much more frequently than a child does. The teachers' signings were mere stimuli rather than significant conversational turns. Summing up the results of this close order work, supplemented by studies of the Gardners' material as well, Terrace remarked '[these analyses provide strong evidence] supporting the idea that prompting had played a much greater role than heretofore recognized in conversation between chimpanzees and humans.'

Several other programmes of like experiments have been set up. For example, other chimpanzees have been taught the use of distinctive counters, a bonobo (a species of small ape, *Pan paniscus*, of the same genus as the chimpanzee, *Pan troglodytes*) has been taught signing by the same complex of methods as that used by the Gardners. All sorts of claims have been made by the experimenters, but the gap between the minds of apes and those of human beings seems as large as ever. However, there is no doubt that these programmes have shown that the difference could be interpreted as one of degree rather than one of kind.

After bringing the Washoe programme to an end Beatrice and Allen Gardner attempted to establish a 'signing culture' among a small group of chimpanzees. An older one, which could use ASL, fostered two infant chimps. Would the infants learn the rudiments of the signings of which the foster parent was capable? The results of this phase of the experiment were apparently inconclusive.

Washoe's subsequent history merited an article in the *New Yorker* describing her lonely life in a chimpanzee colony in which she was the only one capable of signing. Latterly with two of her offspring she had been living at the Chimpanzee and Human Communication Institute at Ellensburg, Washington. In the midst of all this she was involved in an incident in which the distinguished neuroscientist Karl Pribram lost part of a finger. He had put his hand through the metallic boundary fence and she dashed across to warn him of the danger or perhaps for some other reason. Somehow she banged his hand down on the metal chopping off a finger. She died in 2007 at the age of 42.

The human capacity for inflicting pain to order

Reaching out beyond the framework of the logic of the intensive design Stanley Milgram (1933–84) recruited a large number of people at various stages of his well-known experiments on how human beings respond to morally outrageous orders from an apparently properly constituted authority. The hundreds of people who eventually took part were selected by random raids on the local telephone directory. Originally Milgram chose the extensive design, looking for trends across a large number of participants. But as the anomalies in his results began to surface he used post-experimental interviews with participants to move more towards the intensive design, looking for clues as to the type of person who responded in a certain way. However, many of those who were shocked or puzzled by the results of the programme presumed that they could read off these results as having application to all human populations.

The experiments were originally set up to enable the experimenter to choose between two rival hypotheses—a logical pattern which we will encounter in the investigations to be discussed in the next chapter. But as the experiments went on, it became clear that this was an exploration of psychologically unknown territory.

The original question to be addressed by the experiment was this: Are the situation and the social relations within it more or less important than the dispositions of the individual people involved when it comes to accounting for such behaviour as torturing people by giving them painful electric shocks on the orders of a constituted authority?

The attempt to submit the question to experimental test was carried out by Stanley Milgram in the years from 1960 to 1963. The idea of a systematic experimental programme to explore the conditions under which people conformed to the demands of authority came to him in the autumn of 1960. Solomon Asch had already shown that people have a tendency to shift their stated opinions towards those expressed by the majority, but the topic was a simple comparison of the relative length of three lines presented on a screen.[5] If a majority of the viewers declared the lines to be equal length, coached by Asch to collude in this claim, then the one subject left who was not in the know and saw that the lines were unequal tended to change his voiced claim to that of the majority.

Milgram wanted to make the experiment much more realistic where the upshot of the acts a person was ordered to perform had some significance. In his recent biography of Stanley Milgram, *The Man who Shocked the World*, Thomas Blass suggests that Milgram's enthusiasm for and persistence in his 'obedience' studies came in large part from an attempt to understand the Holocaust.[6] How could the appalling behaviour of the thousands of people involved in carrying out the 'Final Solution' be explained?

Thinking of Milgram as the experimenter, an onlooker, and recorder, his apparatus included three living components. There was a person who was introduced as the 'learner'. Then there were the people, who, unbeknownst to themselves, were being experimented upon by Milgram. They were recruited as 'teachers'. Ostensibly the experiment was to test the efficacy of electric shocks to expedite learning. Like the play within the play in *Hamlet*, here was a procedure presented as an experiment within the experiment. The learner was a professional actor who was required to make suitable responses to what the teachers believed were real electric shocks of increasing severity. Every time the learner seemed to make a mistake, the teacher was ordered by

Milgram or his assistants, the third human element in the setup, to increase the voltage. The range of available shocks went from 0 to 450 volts, clearly marked on the console used by the teachers to administer the 'training programme' (Figure 9).

The point of using this complex equipment, of wires, consoles, and people, was in the response, the 'prods', that Milgram or his assistants gave to the protests and worried questions of the teachers: 'Please continue'; 'The experiment requires that you continue'; 'It is absolutely essential that you continue'; 'You have no other choice. You *must* go on.' To doubts expressed about the safety of the experiment Milgram replied, 'Although the shocks may be painful, there is no permanent tissue damage, so please go on.'[7]

Figure 9. The experimental setup. A subject in the obedience experiment at Yale.

Note that Milgram's response sequence has two components—one is an imperative speech-act ordering the teacher to proceed with the 'training'. The other is a putative statement of fact—that the learner, despite appearances, is not being injured by the electricity administered by the teacher. We have two social relations set up by this exchange. One is obedience to authority, the other is trust in the representative of an institution known or believed to be benign. Most of the people who took part as 'components of the apparatus' were tense and nervous—some extremely so (42). When the learner was not visible to the teacher and only his cries of protest were heard, an astonishing 65% of the teachers went on to the lethal limit of 450 volts. When the learner was visible the proportion still held at 40%.

Milgram tried out his experimental procedure under all sorts of conditions and in all sorts of environments. Of course, the basic discursive pattern of command and authoritative reassurance was the same. The problem with every one of these experiments was how to interpret the results. What was the relation that Milgram's procedure established between himself as the experimenter, the person presumably supervising this exercise in teaching, and the true subject of the experiments? In this case one could hardly call the people he recruited 'participants', since they had no idea what was really going on. Reflection on the results and the comments that the 'subjects' made subsequently suggests that the key social psychological relation between the 'teachers' and Milgram was not obedience but trust. At the same time there was the situation itself, interpreted for the subjects as something well worthwhile taking part in.

In short the subjects believed they were doing something morally good. They trusted the words of the person in charge that despite appearances the learner was not being seriously hurt.

This makes the claim that the experiment reveals the psychological inwardness of the Nazi death camps and their guards, operators, and managers deeply suspect. The people who were recruited to run the camps knew very well that the sequence of procedures that led to the gas chambers was a one-way track to death. Dragging the bodies to the incinerators after removing the gold fillings from their teeth could hardly be interpreted in some more benign way. At no point do we have anything comparable to the reassuring 'no tissue damage' responses made by Milgram and his assistants to remonstrations by the teachers.

The analysis does not yet go deep enough. It suggests that about two-thirds of the people recruited by the newspaper advertisement brought to the experiment the same personalities and characters that they left home with. This assumption, widely held not only among lay folk but by psychologists as well, was thrown into doubt by another experiment conducted by Phillip Zimbardo.

Setting up a basement area in Stanford University as a simulation of part of a prison, he recruited students to play the roles of guards and prisoners. They were supposedly instructed in the way to fulfill these roles in accordance with the official code governing this kind of institution. Zimbardo planned to run the experiment for several weeks to see how each group adapted to the roles they were, at first, just pretending to play. Very soon, however, the situation deteriorated as the 'guards' adopted more and more brutal disciplinary measures to which the 'prisoners' retaliated in kind. The experiment was brought to a premature end as the violence escalated. Zimbardo has connected his unexpected results with some of the questions raised by Milgram—how can ordinary people be brought to inflict injury on others?[8] Recent attempts to replicate the original Zimbardo experiment have displayed the sharpness of the role divisions that emerged in the original study. Both guards and prisoners overplayed their roles, perhaps pushing each other to excess.

Years earlier, Erving Goffman had revealed how much people adjust their personalities, opinions, and so on in moving from one social situation to another. He called this phenomenon 'the presentation of self in everyday life', a phrase that later became the title of his well-known book.[9] The activities of the people in social situations are like the performance of a play. The sorts of characters that people display in the situations of everyday life are like parts and roles in the drama unfolding on that stage.

From the point of view of this book, we must carefully delineate the apparatus used in Milgram's experiments. It consisted of the electrical equipment, the people managing the experiment, the subjects, and the professional actor who played the part of the learner. It is something like a circuit with battery, resistors, a Wheatstone Bridge, and so on, which can be used to study the way that a current passes through it. Various instruments like voltmeters and ammeters are located here and there to monitor the flow. In a similar way Milgram's assistants monitored the strength of the electric current that the people thought they were using and their responses to the actions they thought they were performing.

Milgram's experiment was carried out with a human being as the alleged learner who the subjects as teachers were supposed to be training. C. L. Sheridan and R. G. King had the bright idea of substituting an animal for the person in the training situation. The apparatus was essentially the same, except where a human being had slotted into the 'space' for learner now it was a puppy.

The new version of the apparatus changed the moral quality of the overall procedure as seen from the point of view of the psychologists using these two versions of the apparatus to experiment on the people recruited as teachers. In Milgram's version the learner was simulating the effects of imaginary electric shocks. In the Sheridan and King version the puppy had to be presenting real signs of discomfort and suffering since it was actually being given electric shocks. From the point of view of the people filling the teacher slot the psychological/moral situation was the same. Both groups of teachers believed that the responses of the learners were authentic and that they were really administering increasingly painful 'lessons'.

Describing their experiment, Sheridan and King declared,

> In this experiment the learner-victim was actually given shocks. A non-human subject—a cute, fluffy puppy—was substituted for the learner-victim of Milgrim's paradigm. In addition shocks were amperage-limited, and capable of creating responses such as running, howling, and yelping, without however doing the subject any serious harm... The first of the three actual voltage levels produced foot flexion and occasional barks, the second level produced running and vocalization, and the final level resulted in continuous barking and howling.[10]

Sheridan and King found that with their very restricted sample of human beings, 13 men and 13 women, all students, the evident suffering of the 'cute, fluffy puppy', real in this case, brought the 'training procedure' to an end for some of the teachers but not for all. The participants were told to administer up to 30 shocks of increasing severity. All the women went on to the final level at which the puppy was 'continuous[ly] barking and howling'. This came as a great surprise since in those days most people would have agreed with the aphorism 'Girls are Sweet and Neat but Boys are Rough and Tough.' As Sheridan and King put it 'Females [women] were not expected to be more willing than males [men] to shock a cute puppy.' When told to give shocks to a cuddly puppy, boys would be more likely to comply than girls surely—not so! Only 50% of the men in the experiment (7 out of 13) went on to give what they saw were severe shocks to the puppy, while 100% of the girls did when told to do so and when it was made clear that the grade for the course depended on obedient participation in the experiment.

Many of the participants tried to coax the puppy to avoid the shock, while some of the women students became so upset as to begin to cry. It seems obvious that the women were less likely to rebel, though actually more upset. Just as in Milgram's original experiments, those who protested were told firmly to continue, and that it was scientifically important to discover the 'critical fusion frequency' in the vision of dogs, whatever that might be.

As we saw, the predominant theme emerging from Milgram's interviews with the teachers was the trust the teachers had in Milgram and his assistants. Curiously in the light of the result of the Sheridan and King experiment one of the teachers told Milgram that she did not think any of the men who had taken part would have gone on up to the 450-volt level. It seems from the context that men were less likely to take the word of a supposed authority for something than women were.

Yet, the ambiguity in the interpretation of these results is not quite the same for the puppy experiment as it is for the human case. In Milgram's experiment the setting is doubly benign in that the presumption that the staff of universities are to be trusted ties in with the assertion that the project is designed to benefit people. The puppy experiment is simpler and cruder, since the social institutional relations among the various components of the experiment apparatus is stripped down. Furthermore an experiment on canine vision, teaching a dog to discriminate between steady and flickering lights, is hardly an unambiguously worthwhile enterprise.

Commentary

Philosophers have puzzled over the question of the relation between locally obtained items of evidence and the credence they give to related hypotheses. Many have tried to formalize the relationship in terms of ratios of favourable and unfavourable items of evidence to the scope of the generalization they support or undermine. Most of this looks pretty unrealistic when one pays close attention to how scientific research is actually carried out. Over and over again important results are derived from a few or even just one exemplar. This is the intensive design. Checking to see whether the first experiments were done well and the results reliable is rarely the first step to accumulating lots and lots of confirming cases. The Vacanti brothers proved their point with just one pinna riding on the back of just one mouse. The Gardners had Washoe and Terrace had Nim Chimpsky. No more chimps seem to have been needed to sustain a powerful and subtle back and forth between hypotheses and evidence as each programme explored unknown territory. In the end Milgram had to take account of individual responses to the seemingly genuine experimental procedures. Neither those who had professional knowledge of electricity nor members of the clergy went beyond a very modest limit.

Chapter 6
Testing Hypotheses: More Dogs, Monkeys, and Ticks

We can look on the experimental programmes of Stephen Hales and the Oxford cognoscenti as the systematic exploration of further features of the mammalian blood system that could be studied in the light of Harvey's theory of blood circulation. If we assume that the heart is pumping a small quantity of blood round and round the body it is an obvious next step to ask how rapidly it is circulating and what sort of pressure is built up in the system. Hales' programme of experiments was not designed to distinguish between two rival hypotheses about the internal circulation and pressures of the fluid in the mammalian veins and arteries. He was exploring territory opened up by taking the circulation hypothesis for granted. However, many research programmes have used experimental apparatus in order to try to distinguish between two apparently rival hypotheses. Research work has often been advertised as a 'crucial experiment'. Such an experiment would confirm one proposal and disconfirm its rival. This principle would only be viable if we could be sure that when one hypothesis is true the rival one must be false. Of course logicians have pointed out that it is rare for the proof of one hypothesis to refute the relevant other absolutely.

The form of a logical rivalry might look like this: 'All swans are white' is the rival of 'No swans are white.' If one is true, the other must be false. Looking a little more deeply one sees that the best evidence that can be obtained in our earthbound laboratories and ornithological expeditions in the here and now is something of the form 'Some swans are white' since commonsense suggests there might be swans in Australia which have yet to be examined. As a matter of fact they turned out to be black. We can never be sure from experimental evidence that somewhere in the vast reaches of space and time there might not be a life form that is not based on water. We know now that not all life forms depend on oxygen. Some microorganisms consume oxygen and some consume sulphur. So even if all the A's in the here and now turn out to be B's we cannot safely declare on that basis alone that they all have, are, and will be. The discovery of a single A that is not B would be enough to knock out 'All A's are B.' However, it would not go very far in establishing that 'No A's are B.' After all, we already know some are.

The consideration of rival hypotheses can be very complicated. Was the climate the same or different before and after the Flood? Did the hyaena's cave settle the conflict between rival hypotheses? Well, not quite. There never was a Flood.

In this chapter we explore three major experimental programmes that used apparatus based on living things in an attempt to resolve the tension between certain important rival scientific hypotheses. In the first two cases—the nervic and the humoric hypotheses about the control of digestive processes, and the comfort provider and food-giver theories of the origins of mother–infant love—the research programmes were posed as 'crucial experiments' by those most closely allied with them. We will look also at a third piece of scientific research, based on observation rather than experiment, that illustrates how there are qualifications to most attempts to set hypotheses up as exclusive alternatives.

Reshaping dogs: the art of physiological surgery

The experimental programme to which Pavlov contributed so much exemplifies the dual logic of a great deal of experimental science. One can experiment only on what is in the here and now. Yet general and often sweeping conclusions are drawn from the results. Pavlov's dogs, and those that played a key role in the work of Bayliss and Starling, were located in two logical structures. They served as *typical instances in an inductive pattern*, reasoning from the results obtained with this dog or perhaps a few more, to the biology of all dogs. They also served as evidence for more sweeping claims about the biology of all mammals, humans included. This further expansion of the claims for the experiments' results involved the assumption of an essential similarity in digestive systems among hippos, dogs, human beings, and so on.

Ivan Petrovich Pavlov was born on 26 September 1849 in the small town of Ryazan in Russia. When he was nine years old he fell from a fence onto a brick floor landing on his head. After the accident he was taken to live with his uncle, the Abbot of St. Trinity Monastery. There he was encouraged to read very widely and of course to help his famously industrious uncle in household chores. In later life he put down his enthusiasm for the experimental aspect of physiology to the tasks that fell to his lot as a child.

In 1870 he began his studies at St Petersburg University. At this time he was as much filled with the idea of a science-led transformation of society as he was with enthusiasm for scientific knowledge for its own sake. His private reading led him to the Englishman George Lewes's popular works, particularly on biology, long passages from which he learned by heart.

Under Ilya Fadeyevich Tsion (1842–1910), Pavlov carried out a detailed study of the pancreatic nerves, for which he was awarded a gold medal. He had been drawn to the idea of the nervous system as the main means, indeed for a while he believed the only means, by which the internal organs were stimulated to perform their various functions, particularly the secretion of the various digestive juices. Picking up an old conceptual distinction, he distinguished between his 'nervic' theory of the management of the internal organs and the 'humoric' theory of chemical influences. Only later did he extend this idea as the basis of a general neurophysiological approach to psychology, in his studies of the conditioned reflex.

Pavlov enjoyed working with doctoral students on collaborative projects. He said later 'from our discussions I gained the habit of "physiological reasoning" [later to emerge as a distrust of mentalistic explanations]. I progressed until no laboratory technique held any secrets for me.' Visiting Carl Ludwig's laboratory in Leipzig during the years 1884–6 he perfected his skills in 'physiological surgery'.

Pavlov was averse to experimenting on detached parts of animals. He believed that the animal was a complete integrated system with each part influencing all the rest through the activity of the nervous system. His developments of surgical techniques to sample processes, particularly of digestion, were driven by the principle that each phase of the process was related to all the others. He instigated the construction of a laboratory where surgical procedures could be carried out on his dogs under anaesthetic, recreating the dog in the form of an apparatus with portals through which the secretions involved in various phases of the digestive process could be extracted. The point almost always was the relation between one phase of the process, say the presence of food in the stomach, and another phase, say the onset of gastric secretions; then on to the next phase, the beginning of pancreatic secretions. Not only did Pavlov have a horror of blood, so that all his surgical work was conducted with someone cleaning up as the operation unfolded, he insisted that though animals were transformed during physiological surgery, 'with this must go hand in hand the means of healing the injury inseparable from the operation, and of restoring the animal to its normal condition in so far as the nature of the procedure permits.'[1]

This research programme into the nervic control of the digestive process was inspired by William Beaumont's (1785–1853) pioneering studies of the chemistry of digestion. These depended on the accidental fact that an opening or fistula from the stomach to the outside of the body of his Canadian servant, Alexis St Martin, had healed without closing up. It left a permanent opening from which samples of the progress of digestion could be extracted.[2] Pavlov's maturing surgical skills enabled him to prepare experimental dogs in a remarkable way. He created fistulas (external permanent openings) to obtain samples from the salivary glands, the oesophagus, the stomach, the liver, the pancreatic duct, and even the small intestines. This enabled him to follow the process of digestion in extraordinary detail, particularly to make exact measurements of the quantitative relationships between stimuli and gastric responses. At first Pavlov's preparations of fistulas in the bodies of dogs with his techniques of physiological surgery quickly closed up as they healed. He solved the problem of keeping them open by the ingenious idea of sewing a small piece of mucous membrane into the lips of the fistula, mimicking the mouth, nostrils, and other external openings of the body. The fistulas he created were permanent.

His work was attracting considerable international attention. In 1904 he was awarded the Nobel Prize for his researches into the neural mechanism by which the secretion of gastric juices was stimulated. However, in his Nobel speech he spoke a good deal on his recently begun investigation of the conditioned reflex. It was this work which was to prove so influential in psychology and for which his name is commonly known.

Eventually he was appointed to the Chair of Physiology in the Military-Medical Academy. Shortly afterwards he took on the direction of the physiology division of the Imperial Institute of Experimental Medicine.

After the revolution of 1917, he gave his general but critical support to the new regime. In so far as it had emphasized the advancement of science it had his full commitment. However, he seems to have stood back from the 'social engineering' of the early Bolshevik regime that eventually went so tragically wrong. Over the years he had moved away from the dogmatic materialism of his youth, though he never ceased to be vigorously opposed to any idea of the mind as a mental substance, separate and detachable from the body.

A great many dogs were sacrificed in Pavlov's pursuit of knowledge (Figure 10). For example, one of his students reports that to run the 'gastric-juice factory', the gastric juice was collected from six to eight dogs which had undergone a surgical procedure that detached the oesophagus from the stomach. A gastric fistula led

Figure 10. Pavlov with one of his dogs.

out from the gullet. The dogs were given minced beef to eat 'which because of the severance of the oesophagus fell out from the upper end of the gullet onto a dish—and the dogs gobbled it up anew.'[3] By this method a large quantity of gastric juice can be collected from fistulas that open out from the stomach. It is not clear how any of the dogs could survive this procedure. In a speech by an admiring colleague, V. O. Samoïlov, Pavlov was congratulated for the perfection of the technique for creating a 'gastric pocket', a portion of the stomach active yet fenced off from food: 'thirty dogs were sacrificed in vain—much time—almost six months—and much effort, were expended [before success was achieved].'

The experiment we will focus on was a test of the nervic hypothesis of the control of the digestive secretions. Pavlov had become convinced that the links between the mouth, the stomach, and the pancreas involved an important bundle of neural fibres, the vagus nerve. Some years earlier Sir William Sherrington had demonstrated that nerve fibres had both excitatory and inhibitory functions throughout the system of motor control and beyond. Pavlov realized that this should also be true of vagus nerve fibres. In a second phase of the experiment he set about demonstrating this dual control of the gastric and pancreatic secretions, exciting and inhibiting them.

To follow the control of the digestive process food must pass from the mouth to the stomach and on to the intestines. At each stage digestive substances are mixed with the food, 'juices' secreted by the local glands: saliva in the mouth, hydrochloric acid in the stomach, and pancreatic juices and bile in the duodenum. But if these secretions are to be collected in a pure enough form to be chemically analysed they must be free from contamination by the food. The solution was the Pavlov pouch. An incision in the stomach wall allowed Pavlov to create two flaps of tissue which could be sewn into a pouch, with a seam on the inside separating it from the rest of the stomach. An external fistula was constructed from which uncontaminated secretions from inside the pouch could be collected. The preparation of the pouch was a very skilled matter since Pavlov needed to preserve intact the fibres of the vagus nerve bundle as they passed through the stomach wall, if the pouch secretions were to be relevant to the testing of the nervic hypothesis. A great deal of work was needed to perfect the operation.

To give the flavour of Pavlov's style of scientific work, with its mixture of surgical skill, insight, and showmanship, here is a transcript of the part of his 1897 lecture series on the digestive glands that includes this experiment. Commenting on previous studies as he proceeded with an experimental demonstration Pavlov began his lecture as follows:

> It resulted that the vagus [nerve bundle] is to be looked upon as the secretory nerve of the pancreas. For this discovery we have to thank the adoption of a special surgical procedure . . . the dog before you is provided with a permanent pancreatic fistula . . . four days ago the cervical vagus was divided on one side; the peripheral end of the nerve was laid bare, furnished with a ligature, and preserved under the skin. [Addressing the audience he goes on] I now carefully remove the cutaneous sutures, and continuously draw forward the ligature with the nerve, without causing the dog appreciable discomfort. I beg you to notice that from the metallic funnel, the wide end of which embraces the site on the abdominal wall where the orifice of the pancreatic duct is situated not a drop of juice flows. I now begin to excite the nerve with an induction current . . . the dog remains perfectly still . . . two minutes elapse without any result from the stimulus . . . and now in the third minute the first drop of juice makes its appearance and is followed by others in quicker and quicker succession . . . [after the stimulus is interrupted] the juice continues to flow spontaneously . . . [for four to five minutes] . . . I again apply the current, and obtain the same effect invariably upon every dog.[4]

The last sentence of the extract brings to light the logical framework of inductive reasoning which is one paradigm within which to set these experimental results. At the same time one must bear in mind that the title of lecture series was 'The Working of the Digestive Glands' without explicit restriction of time, place, or species.

The final step in the programme involved the testing of the idea that the vagus nerve carried both excitatory and inhibitory messages. Another study had shown that stimulating the pancreas independently produced copious secretions but these continued for as much as an hour. In an intact body something must halt the process. By separating the excitatory from the inhibitory pathways in the vagus nerve, and stimulating first the excitatory and then the inhibitory fibres, the secretion was stopped, at least for a while. Pavlov took it for granted that the nervous stimulus acted directly on the pancreas, rather than through a chemical intermediary.

Even so, the delays between electrical stimulation and the onset of glandular secretions remained unaccounted for. This was finally to be explained by the rival humoric hypothesis of William Bayliss and Ernest Starling—the idea that chemical stimuli were involved and perhaps exerted the paramount influence on digestive secretions.

The procedure used by Bayliss and Starling had already been used by Pavlov and was described in the same lecture series. The physiological surgery was much more drastic than in the original experiment. The spinal cord, trachea, and oesophagus were severed and the dog kept alive by artificial respiration. Opening the chest Pavlov isolated the vagus nerve. By electrically stimulating this nerve he obtained secretions from the pancreas. This was the procedure from which Bayliss and Starling began.

William Bayliss (1860–1924) managed to combine an active life in business with a distinguished career as a physiologist. After studying physiology at Wadham College, Oxford he joined University College, London as a physiologist. He soon teamed up with Ernest Starling, and was involved in a wide range of physiological research.

Ernest Starling (1866–1927) entered Guys Hospital in 1882. During his medical training he worked closely with Leonard Woolbridge on the physiology of the heart. After Woolbridge died Starling not only took over his job but married his widow, whose considerable fortune enabled him to continue his researches without taking up a medical practice. In 1899 he moved to University College, London where he met William Bayliss.

The two physiologists seem to have had one of those typical partnerships that flourish in the sciences—Bayliss the quiet one and Starling brisk, opinionated, and ambitious, rather like the steady Morley and the excitable Michelson in physics. They were close personally. Bayliss married Starling's sister.

Generally Pavlov's physiological surgery reshaped a complete organism into an experimental apparatus, even in the experimental procedure described earlier from which an intact animal could not possibly have recovered. Bayliss and Starling abstracted a portion of the animal for the purposes of the experiment, much as a chemist might run a reaction in a single test tube. The following is a paraphrase of their account of the experimental procedure.[5]

'On January 18th, 1902, a bitch of about 6 kilos weight, which had been fed 18 hours previously, was given a hypodermic injection of morphia 3 hours before the experiment.' Just as Pavlov had done they isolated the vagus nerve bundle, but cut it effectively, isolating the stomach and intestines from any stimuli from the nervous system. A loop of the intestine immediately following the duodenum was tied at both ends and the nerves supplying this region 'were dissected out and divided, so that the piece of intestine was connected to the body of the animal merely by its arteries and veins'. By inserting a tube into the large pancreatic duct they were able to record the secretion from the gland.

Food that moves from the stomach to the duodenum contains hydrochloric acid. When Bayliss and Starling introduced dilute hydrochloric acid into the duodenum the pancreatic secretion began. However, the pancreatic secretion was also produced by addition of dilute hydrochloric acid to the completely enervated section of the jejunum. It had already been shown that introducing hydrochloric acid into the blood stream had no effect on pancreatic secretions. 'The conclusion was inevitable', Bayliss and Starling declared, 'that the effect was produced by some [other] chemical substance finding its way into the veins of the loop of jejunum in question and being carried in the blood stream to the pancreatic cells.' What was this mysterious substance? It must have been produced by the cells lining the wall of the jejunum.

If the action of the acid on these cells produced the 'body' they were in search of, the next step was clear—'to cut out the loop of jejunum, scrape off the mucous membrane, rub it up with sand and 4% HCl in a mortar, filter through cotton wool to get rid of the lumps and the sand, and inject the extract into a vein'. The first effect was a drop in blood pressure which they subsequently showed was due to another chemically distinct substance. Then, after a short time a substantial flow of pancreatic juice began. They named the active substance 'secretin'.

In reworking some of Pavlov's researches they had prepared a purified extract of the lining of the duodenum. This stimulated a strong secretion of 'pancreatic juice', without any contributions from the nerves. It must have contained a *chemical agent* powerful enough to activate the pancreas. In the old terminology it was humoric.

The publication of the work of Bayliss and Starling on the role of secretin in the stimulation of pancreatic secretions in 1902 changed Pavlov's mind about the exclusively nervic mechanisms of the control of digestion, but not without a struggle. Here was proof positive of the humoric or biochemical character of at least an important part of the control path of the sequence of digestive processes.

When he heard of this result Pavlov instructed V. V. Savich to repeat the study. 'Pavlov and the rest of us watched the experiment in silence. Then, without a word, Pavlov disappeared into his study. Babkin reports that he returned half an hour later and said, "Of course, they are right. It is clear that we did not take out an exclusive patent for the discovery of truth."' The Russians were convinced that neural mechanisms still played a role. Pavlov and his colleagues adopted a kind of hybrid theory—involving both nervic and humoral influences. Bayliss and Starling initially went the other way. They denied that the vagus nerve had any role in the stimulation of the pancreas. Eventually the hybrid theory prevailed in both camps. In the course of these researches and controversies Starling popularized the word 'hormone' as a general name for 'chemical messengers'.

The physiological surgery needed to create the apparatus for the Bayliss and Starling experiment effectively reduced the dog to a collection of parts, which could not be reassembled into a whole dog again. Sometimes this was true in Pavlov's laboratory, but not always. Babkin tells of dogs living for years with the fistulas that had been created for experimental purposes.

Strongly and publicly opposed to 'crude vivisection' of animals, Pavlov was a foremost defender of their humane use in scientific studies. By that he meant the surgical reshaping of the bodies of his dogs under anaesthetic into pieces of scientific apparatus. He set up a memorial to his dogs on which he put the following inscription:

> The dog, man's helper and friend from prehistoric times, may justly be offered as a sacrifice to science, but this should always be done without unnecessary suffering.

Thanks to a generous donation from Alfred Nobel in 1893, Pavlov was able to erect a purpose-built set of laboratories in which to carry on the researches he had laboured to perform under the straitened conditions of his early years. He died in 1936, still active in neurophysiological research. The last phase of his life was devoted to the exploration of the conditioned reflex, which he took to be the foundation of all seemingly mental activities.

Cupboard love or cuddles?

For many years a member of the Psychology Department at University of Wisconsin, Madison, Harry Harlow undertook a long-running series of studies to explore the conditions under which maternal and filial love developed or failed to develop with young monkeys. There were two seemingly contrary hypotheses in play: did filial love arise from the provision of food or did it arise from the provision of bodily comfort? These studies could also be seen as a way of testing John Bowlby's famous theses about maternal attachment and loss, and the effect of filial relationships and their disruption on children. The question Harlow addressed was which of two simple alternatives was correct: is the relation between a youngster and its mother forged by a supply of food or by the provision of comfort?

Common observation suggested that interruption of a close relationship between mother and infant human beings had some long-term effects on the infant, not always for the best. Could Harlow find a way to create a situation which would be equally damaging to or productive of the mental health of infant monkeys in a laboratory under apparently controlled conditions? That is just what he did!

The apparatus that Harlow created has a ghoulish look that has made it irresistible as an illustration to commentaries both favourable and hostile to this project and the way it was carried out. Importantly the apparatus consisted of an intimate blend of inorganic and organic or living components. A bare wire frame offered a source of food while a wooden frame covered in a towel provided comfort as a surrogate mother (Figure 11). Both were offered as alternative resting places for a living monkey infant.

There have been some very searching criticisms of Harlow's methods. Some of them have focused on the moral justification for this work. It is not my intention in this book to go into those matters. My purpose is to describe as clearly as possible how apparatus in which animal/plant material had an essential role were constructed and used, together with something about the experimenter who used it.

It has been said of Harlow that he was a person who, though a great success as scientist, was a tragic failure in life. While we can have little reason to doubt the truth of the second clause of this remark, one must be sceptical of the first clause. Harlow's scientific work has not stood the test of time. To understand why it had such an impact on 1950s America we must see it in the context of behaviourism's dominance in the country, and particularly the version inspired by B. F. Skinner. In the USA behavioural therapies and token economies with which good behaviour was rewarded dominated remedial therapies. The idea that mothering practices might have a permanent effect on the psyches of children was disturbing. In Europe the results of Harlow's experiment produced very little of a stir. Freudian theories covered much the same ground. More importantly, in the context of therapeutic practice in dealing with the problems of troubled children the work of John Bowlby springing from his studies of real children under real conditions of deprivation and loss were sufficiently definitive to need no support from tenuous generalizations from the lives of laboratory monkeys.[6]

Figure 11. A surrogate monkey mother.

I got to know Harry Harlow a little when I was visiting the University of Wisconsin at Madison in 1965–6. His flat voice and his unwillingness to engage in anything resembling a conversation did not make visits to his lab easy. The wire cages and the 'clinical' atmosphere in which the hang-dog subjects of his experiments sat and moped did not help either. His general resemblance to Senator Eugene McCarthy was disconcerting.

Fortunately for us Deborah Blum has written a lively and detailed biography of Harlow, which includes very full accounts of his experimental programme.[7] Perhaps more importantly, it also includes a thorough study of the life of this unhappy and even tragic person.

The research interests of psychologists all too often seem to be dictated by problems that they face in their own lives. Henri Tajfel, well known to social psychologists, moving from his native Poland to France and then to England, spent his life struggling to understand how people came to form social identities. What was an artefact for him was 'second nature' for most of us. Harry Harlow seems to have been tied to a similar treadmill. His life represents a sad story of someone for whom the love and affection of other people was always problematic, even from his lonely childhood. His first marriage ended in divorce and it is said that he became estranged from his sons. His second marriage, to one of his assistants, was cut short by her early death from cancer. When I knew him his descent into alcoholism had already begun.

Harry Frederick Israel was born in the small country town of Fairfield, Iowa, on 31 October 1905. He was the third of four brothers. When he was still an infant his older sibling was diagnosed with tuberculosis of the spine. Harry's mother, as one would expect, devoted a great deal of her time and affection to the invalid. Indeed the life of the family was focused for many years around this misfortune. Harry's father, Alonzo Harlow Israel, seems to have been a somewhat feckless character, drifting from one failed project to another. Moreover, he abandoned his only viable enterprise, a country store, to take the whole family to New Mexico for the sake of his second son's health. After various disasters the family returned to Iowa and opened another country store, which provided a modest but secure life.

Harry Israel went to the local high school. From all accounts he was a solitary child without much in the way of the usual companionship of other children his own age. The Israel parents were like many people of the era, determined that their sons should have a first-rate education. After a year at Reed College in Oregon, Harry and his eldest brother entered Stanford University in 1924. After a short time he changed his course from English to psychology, though he had some talent in literary matters. Even then he displayed a not uncommon mix of personal shyness and argumentative aggression.

The leading figures in the Psychology Department at Stanford at that time were Calvin P. Stone (1892–1954) and Lewis M. Terman (1877–1956). Harry was soon working under the immediate supervision of Stone. Looking back we sometimes attribute the crude positivism of mid-century American psychology to the influence of J. B. Watson and B. F. Skinner. However, this kind of 'scientism' was very widespread, not least at Stanford. Stone believed that the royal road to psychological understanding was work with animals, and especially rats. Simple experiments based on the methodology of dependent and independent variables would suffice to reveal the intricacies of human psychology as well. What is more, the welfare of these creatures could justly be sacrificed on the altar of 'science'. In one of his experiments on the source of maternal behaviour in rats, carried out by Israel under instruction from Stone, female rats were subject to various manipulations. Some were blinded, in others their brains were dissected to destroy the sense of smell, all to see whether they would still strive to recover their pups even under so diminished a condition. Harlow later confessed a particular dislike of 'rodentology'. In his own practice he was perhaps too easily satisfied with 'simiantology'.

The head of the department at Stanford, Lewis Terman, took a great interest in establishing the careers of the graduands, in particular getting them promising jobs. Harry Israel was offered a job at the University of Wisconsin at Madison. However, Terman was convinced that the latent anti-Semitism of the era would count against the new assistant professor. He persuaded Harry to change his name to Harlow, even on the title page of his dissertation. This seems an extraordinary story—particularly if one remembers that in that university there were a great many distinguished professors of Jewish origin such as Julius Weinberg in Philosophy and Leonard Berkowitz in Psychology.

There Harry Harlow remained for the rest of his troubled life. He had been a lonely child, and from Blum's account he contrived to become a lonely adult. Immersing himself in his work he could distance himself from family life. The solace of alcohol became important to him, while he single-mindedly forged on with his experimental programme. His general mental state deteriorated markedly in the summer of 1981. He had been suffering from Parkinson's Disease for some time and his isolation acerbated his conditions. He moved to a retirement home where he died on 8 December of the same year.

HARLOW'S METHODOLOGY

Whatever our contemporary judgements of the value of Harlow's work, his experiments using hybrid apparatus, half-nonorganic half-organic, were a sensation at the time. He still has admirers, though there are few now who would laud his data as providing a 'scientific basis' for conclusions that run counter to or support closely focused realistic studies such as Bowlby's observations of the psychological effects of maternal attachment and loss. Of course, calling Harlow's results 'scientific' is problematic. 'Controlled laboratory conditions' sometimes bring forth artefacts. Apparatus does not always succeed in modelling the phenomena it is supposed to clarify. Not least among the problems raised by the attempts to generalize Harlow's work to human beings is the role of the family in all aspects of human development. In the mothering experiments Harlow's monkeys not only had no real mother, they had no family either. In this respect the work suffers by comparison with the real-world investigations of Jane Goodall and other primatologists in which young chimpanzees are closely observed as they grow up in the care and company of family members. Harlow did carry out another series of experiments independently of the surrogate mother studies, in which infant monkeys were deprived of various aspects of normal simian family life. Various incurable social and psychological defects emerged in the animals as they entered adult life.

The work carried out in Harlow's numerous experiments can be represented as if it were an instance of a simple logical paradigm—either maternal–infant bonding is the result of the provision of comfort or it's the result of the provision of nourishment. The experiments are purported to have shown that only one of these is true: comfort trumps nourishment.

At first sight it seems as though that claim is well supported by the experiments. However, close examination of Harlow's studies shows that the actual methodology was based on a quite different logical structure. The case is a little different from that set out schematically in the introduction to this chapter. Suppose we are presented with two hypotheses: (1) 'All X's are Y' ('all filial love is based on nutrition') and (2) 'All X's are Z' ('all filial love is based on comfort'). Harlow found support for the second hypothesis. Only when 'Y' and 'Z' exclude one another does the evidence refute the first hypothesis.

We need to pay attention to the equipment Harlow created to test the hypotheses about maternal–infant bonds. He constructed pieces of apparatus in which only one aspect of infant–parent interaction—abstracted from all other aspects of mothering—was represented. In the end the target of research was the human world of mothers and infants, such as that explored by John Bowlby in his studies of attachment and loss or hypothesized by Freud in his studies of neuroses. So the results of the experiments on monkey mother–infant pairs were severally supposed to be inductive support for a generalization to human beings, supposed in relevant respects to be similar. What would happen if children were fed but deprived of affectionate mothering? Under the other condition comfort would be readily available sometimes with food and sometimes without. These are supposedly representations of family life with a cold but efficient mother and a warm but indigent mother. Or something like that.

The pattern involved two stages of generalization:

1) The wire and towel constructions supposedly represented aspects of monkey mothers to which real infant monkeys were introduced. The first generalization was from the apparatus, in which material things were combined with living beings, to real monkey families in which both constituents were living beings. The experimental 'mothers'

were created by stripping down reality to a representation of the source of nourishment, bottle and teat on a wire frame or to the source of comfort, warm cloth covering on a wooden frame.

2) The second generalization led from the stripped-down monkey world to the human world. One could think of this reasoning as based on the assumption of a second string of similarities ignoring differences. What has been shown to be true for monkey mothering will also hold true for human mothering. So long as the similarities between people and monkeys outweighed the differences, the results of the monkey experiments could be transported across the two inductive bridges from Harlow's hybrid pieces of apparatus to the real world of monkeys and from that to the human case.

Running these models is supposed to provide information about the roots of mother–child relationships in the human world, in so far as the development of emotional ties between mothers and infants is the salient issue. Harlow presented the research as a 'crucial experiment'. If baby monkeys choose comfort then that is what they prefer. That they do not choose the nutritional surrogate shows that the hypothesis that monkey love is based on provision of nutrition is false.

Methodologically these experiments were examples of a different research paradigm to be illustrated in the next section (see below, Encephalitis and the ticks). Harlow had actually created model worlds, analogues of the real world of monkey life forms. They were stripped-down versions of reality in various ways. A model is created by retaining some key similarities between model and source and introducing some differences. Running a model world is supposed to show how a real-world version of the model world would be in certain salient features.

THE EXPERIMENTAL PROGRAMME

Harlow described his experimental programme and presented the results in some detail in a small book that summed up the work of many years, most of which had been published piecemeal in articles.[8] It is important in assessing Harlow's work to take note of the rhetorical presentation of his results as an account of the nature and origins of 'love'. According to Harlow there are a number of 'affectional systems', five in all.[9] Maternal love is tied in with the complementary system of infant love. The third system underlies relations between immature members of a primate species. The fourth system is the foundation of heterosexual relationships and the fifth, paternal love. The uses of the experimental apparatus to be described here were attempts to explore the first and second systems, maternal love for the infant and infant love for the mother.

Harlow's studies depended on the general principle that relations between mothers and infants in the very early period of infancy must be the result of the activation of inherited propensities to follow out certain routines. Konrad Lorenz's influence is clear in this work. The target against which Harlow directed these experiments was the assumption that only some very primitive biological drives were inherent in the organism, such as hunger and thirst. None of these could remotely be identified with love. However, another aspect of mother–infant interactions is certainly evident in actual ways of living, namely the cling–comfort pattern. The experiments with surrogate mothers were meant to try to establish the latter as paramount over the former.

In the quaint rhetoric of psychological reports Harlow describes his work in the third person: 'To determine whether infant love is learned or whether certain inherent properties of the mother elicit infant attachment Harlow constructed sets of nursing and non-nursing cloth and wire surrogate mothers.'[10]

These studies made use of infant rhesus monkeys as the organic component in the apparatus. The surrogates had easily grasped bodies and schematic monkey faces. Some had a single easily accessible breast, while others had none. The total apparatus, variants of which I believe should be construed as various model worlds, consisted of a surrogate mother in an enclosure accessible from the infant's home cage. Four monkey infants were given access to 'lactating' cloth mother surrogates and non-lactating wire surrogates. In the matching experiment the wire mothers 'lactated' and the cloth mothers did not. So what happened?

Very soon the infants which had been suckled by the wire began to spend more and more time clinging to the cloth mother. The graphical representation of the contrasting behaviour of the two groups of monkey infants shows this with startling clarity.

Harlow's conclusion is robust. This is a 'finding completely contrary to the interpretation that infant love is a derived drive in which the mother's face or form becomes a condition for hunger or thirst reduction.' The data show that 'contact comfort' is the main source of infant love.

A second phase using the same hybrid apparatus—cloth surrogate mother and infant monkey—compared the time infants spent on those that lactated versus time spent on those that did not. It comes as no surprise to learn that the combination cloth–lactation was preferred to 'cloth–no lactation'.

What about temperature? Warm wire surrogates were preferred to cold cloth mothers at the beginning. After twenty days, clinging to the cloth mother began to dominate the time spent in all surrogate maternal interactions. Evidently the provision of a warm resting place is less ultimately rewarding than one of a more appealing and more realistic 'monkey' texture.

Harlow's overall conclusion was markedly anti-Freudian. Infant emotional attachments are not egocentric in essence, but are truly interpersonal.

The pursuit of the understanding of love that obsessed this lonely scientist continued into various other dimensions, testing the effect of various kinds and dimensions of deprivation of interpersonal relations on the character and dispositions of adult monkeys. It must be reported that some of these experimental manipulations led to the development of adult characters and capabilities so seriously damaged that a normal rhesus life was impossible for them.

The discussion in Harlow's book moves freely between the worlds of rhesus monkeys and macaques as models for the human world, and that world itself. Similarities and differences are drawn at every point. The first stage, the relation between the apparatus and the relevant real-world relationship among monkeys, does not get the same detailed attention.

Why construct these pieces of apparatus? Why not spend time 'hanging out' with monkeys, as Jane Goodall did with her chimpanzees?[11] There seem to be two reasons, only one of which is strictly scientific. As Harlow notes here and there the apparatus allows representative aspects of mother–infant interactions to be readily studied by abstracting them from a very complex reality. Then choosing surrogate mothers and infant monkey as the experimental apparatus abstracts from the greater complexities of human life.

Encephalitis and the ticks

So far the experimental programmes described in this chapter have looked, at least at first, as if they were examples of 'either this or that' reasoning, the format of the 'crucial experiment'. Later qualifications of the starkness of the alternatives still leaves the distinctions on which the programmes were based intact. It is just that the contrast is weakened. In the case of Pavlov's nervic theory, there are also humoric or chemical stimuli of gastric secretions as well. These two theories exhaust the alternatives. Conveniently both are true. There are both electrical and chemical stimulations of digestive secretions. Harlow's experiments may have exaggerated the contrast between the two poles of mothering, with each represented by its own sharply characterized model. However, in the world of real monkey mothers warmth, nourishment, and a pleasant texture are on offer. Harlow's experiments revealed a preference rather than all-or-nothing alternatives. In the study now to be described the first impression of sharply contrasted alternative explanations of the spread of encephalitis is quickly dissipated as the complexity of the conditions involved in an epidemic become clear.

The studies reported here were carried out by Sarah Randolph and Dana Sumilo of the Department of Zoology at Oxford University. At first sight they appear to be a test of the common-sense suggestion that as winters shorten so the time of year at which tick larvae are produced gets earlier and earlier. Thus the tick population increases and so should the incidence of tick-borne diseases. Theoretical support for this hypothesis seemed to be cast iron, since global warming occurred uniformly across the Baltic States in the early 1990s and tick-based encephalitis (TBE) increased in Estonia and Latvia. In Lithuania the increase was extraordinary—1,065 per cent. If the tick population increased uniformly across the region the incidence of TBE should also be uniform.

Close study of the disease pattern district by district showed marked variations within the overall pattern, though the increase in the incidence of ticks was more or less the same across all three Baltic countries. The 'tick detector' served as a disconfirmer of the simple climatic hypothesis. If ticks had been the main link between climate and disease incidence then there should have been no local variations. Sarah Randolph, the leader of the research project, offered a cluster of alternative explanations, any or all of which might turn out to be salient in further studies: 'Our research suggests that, while changes in climate may play a role, socio-economic effects—such as those of the break-up of the Soviet Union—have a much greater influence.'[12]

The reasoning behind the use of the distribution of ticks to check climate change hypotheses and their relation to the spread of diseases is complex. There are four players in the pattern shown up by the researchers. There are the ticks, the tick-borne disease TBE, a form of encephalitis (inflammation of the surface of the brain), climate change, and the socio-economic disruptions during the collapse of the Soviet Union. All three Baltic States were part of it. The hypothesis to be tested is an alleged simple correlation between climate change in the Baltic States and the upsurge of tick-borne diseases. Ticks, so it was supposed, mediated between the climate and the epidemic. The study showed that while the simple common-sense hypothesis was almost certainly not true, the logical space into which a rival should fit was unoccupied, or perhaps occupied by too many alternatives to sustain even a preliminary 'critical experiment' methodology. If it wasn't the plague of ticks consequent upon the longer breeding season that warmer winter made possible, what could it be?

Showing that tick numbers cannot account for all the details of the epidemic does not dispose of the thesis that tick numbers are relevant to the onset of encephalitis epidemics. It shows that the simple logical opposition 'All or None' does not fit this research. It shows something about the limits of a method that relies upon the abstraction of single factors from a complex background. In laboratory science the insulation and isolation of the apparatus from extraneous influences is supposed to take care of this problem—but in the fields and forests of the Baltic States we are far from the conditions of the idealized experimental setups in laboratories. Even there, few occasions on which the stark alternatives of the 'all or none' pattern are to be found.

Commentary

Within the framework of biology as Pavlov knew it at the beginning of the twentieth century, digestive secretions were either the result of nervous impulses from mouth to stomach, duodenum to pancreas, and so on, or the result of chemical diffusions through the blood system. The only possible mechanisms for the control of digestion were, he thought, either nervic (electrical) or humoric (chemical). Ingeniously reconstructed dogs became the apparatus for Pavlov. His experimental procedure abstracted a neural connection between parts of the digestive system, while Bayliss and Starling, leaching out the active ingredients in the lining of the duodenum, abstracted a chemical connection. How much of the pancreatic secretion could be put down to each kind of influence, all, none, or just some? Did the neural stimulus play a part in the production of secretin, the chemical stimulus to pancreatic activity? Did Bayliss and Starling's discovery of a chemical link between the stages of digestive secretion knock out Pavlov's rival nervic hypothesis? At first it seemed so. In the end it turned out that the workings of the system depends on both kinds of links.

A similar logic constrained the experiments on the basis of mother–infant bonds, undertaken by Harry F. Harlow. He created various mixes of the living and the non-living to create the equipment he needed to run his experiments. Apparatus, half-nonorganic, half-organic, was used to compare the viability of two seemingly very different and apparently mutually exclusive hypotheses concerning the basis of affection between mothers and infants. Was it based on the provision of nourishment or on the provision of comfort? In the end the reader of Harlow's little book describing his researches is ready to conclude 'Both!'

Sarah Randolph and her team showed that the hypothesis that the increase in encephalitis was exclusively due to infection from the bites of ticks in epidemic proportions due to a longer breeding season was not well supported. Nevertheless her observations did not show that it was caused by some other 'vector' or carrier of the infection. There were all sorts of other factors that might have played a role in the way encephalitis infections spread through the Baltic States in the 1990s modifying the simple relation between the prevalence of ticks and severity of the epidemic.

Chapter 7
Modelling Individuals: Yet More Dogs, A Few Mice, and One Human Stomach

A great deal of research in which animal bodies have figured has involved treating them as models of human bodies. The patterns of reasoning are typical of model-based research. In such work, similarities between model and subject must be weighed against differences, not only in their behaviour, but also in their intrinsic natures.

Scientific research is traditionally supposed to be a cluster of practices distinct from engineering, pharmacology, and psychiatry. Scientists, it is declared, pursue knowledge for its own sake. Engineers, pharmacologists, surgeons, and psychiatrists use scientific knowledge to create machines, drugs, and clinical procedures to carry out certain practical tasks, advertised as being of benefit to human beings, their pets, and those aspects of the environment significant to them. Structural mechanics underlies bridge design; organic chemistry underlies the pharmaceutical industry; and psychology is at least supposed to underlie psychiatric practice. However, the distinction becomes fuzzy in certain cases. It does sometimes happen that the development of a practical project brings to light new knowledge.

Engineers and others in the 'applied sciences' business usually begin a project with the construction of a 'pilot plant'. Pharmacologists usually use a sample of a new preparation and test it on populations of subjects, usually by the double-blind method; that is, neither the experimenters nor the patients know who has been given the drug and who has been given a placebo. Sometimes running the plant and carrying out clinical trials adds to the stock of knowledge in physics, chemistry, and pharmacology on the basis of which the project was originally conceived and the test equipment created. A pilot plant is a kind of apparatus and a clinical trial is a kind of experiment with the subjects or patients as apparatus. The logic defines the same framework as that in which model-making and experimenting are done in the sciences as we ordinarily know them.

What if there are not only practical but moral problems with experimenting on the real thing? At one time dissection of the human body to learn about its internal structure was forbidden on religious grounds. In our time, apart from a few scandalous examples, experimenting on human bodies has been very strictly controlled. When the results of a procedure are uncertain, animal bodies have been used as experimental apparatus, based on the familiar principles of modelling. By creating an apparatus sufficiently similar to an inaccessible subject, processes within that subject can be studied indirectly by experimenting with the model. In the work to be described the subject, the human body, was inaccessible for experimentation because it was deemed to be off limits morally. Experimental tests of uncertain heart transplantation procedures could not be performed on human beings. A model or analogue had to be found that was not at that time and place intrinsically morally protected.

If the person whose body is to serve as an experimental apparatus agrees to being used in a scientific research project, the presumption is that he or she gave their informed consent. While there are very complicated issues around the principle of 'informed consent', these are part of the province of moral philosophy and medical ethics, and not within the scope of this book.

In the first section of this chapter I will describe the use of dogs as pilot plants for certain kinds of surgical procedures. These procedures are on the hazy borderline between science and medicine. Much has been learned about the immune systems of humans and other animals, at the same time as much has been perfected in surgical techniques. The procedures we will look at are not just medical, but contribute, sometimes indirectly, to scientific knowledge. Dogs have figured largely as models for the anatomy and physiology of human beings. Later in the chapter mice become important, particularly when the modelling is aimed at understanding how biochemical aspects of organisms are related to psychological phenomena such as depression and aggression. The last example in this chapter concerns a rare occurrence indeed, when a human being deliberately chooses himself as a model in which to run an experiment.[1] This is the story of how Barry Marshall infected himself with the helicobacter micro-organism to establish a model in which to run a test of the hypothesis that he and Robin Warren had proposed as an explanation of ulceration of the stomach. Medical advances came from the proof that their hypothesis was correct, given that the Marshall stomach was sufficiently similar to all other human stomachs, that is, an adequate model to represent them all.

The experiment itself was not aimed at finding a cure for stomach ulcers.

Barnard's baboon

Kidney dialysis machines can perform the functions of living kidneys remarkably well. The same is not true of the liver, where hugely complex patterns of chemical and physical processes are carried on. It was well known that the liver has remarkable powers of self-repair, if it can be rested. No conceivable machine could do the work of a human liver. The analogue between kidney dialysis and liver substitution does not go through. However, if the liver of another species of animal was sufficiently similar to the human liver, and differed in inessential ways, the positive analogy would outweigh the negative analogy. The animal liver could then be not only a model for the human liver but at least a temporary surrogate.

Christiaan Barnard (1922–2001), later famous for performing the first heart transplant operation, had the idea of exploiting the modelling relation between the liver of a baboon and that of a human being. The baboon liver could serve a practical role as well as its possible experimental study as a model. It could be a surrogate organ, sufficiently functionally similar to be linked up to the blood system of a human being, while the patient's liver was 'resting'. During his last week of a visit to the medical school at Richmond, Virginia, he suggested the idea to his colleagues. To the objection that the baboon liver would be 'hostile' to human blood, that is, that there was a serious negative analogy in the way of establishing the baboon liver in a modelling role, Barnard replied, 'I'll take a baboon, and cool him down, wash out his blood [to get rid of natural antibodies against human blood] with water, then fill him up with human blood of the same group as ... [the] patient.'[2]

Baboons were in short supply, but with a sense of urgency, Barnard arranged for one to be air-freighted to Richmond. Hooking it up to a suitable machine he went ahead.

> When the body temperature reached 5C—a deep enough hypothermia to prevent brain damage during the period when the animal would be without circulation or blood—I cut the pump and quickly drained the baboon of all its blood.
>
> It was then hooked up to a second machine, containing Ringer's lactate solution. This washed through until it came out crystal-clear, indicating no blood was left. We were then at a critical point, and I quickly attached leads to the third machine, containing human blood.[3]

On his way to the airport the next morning Barnard called in to see the baboon and found him sitting up and eating an orange. Unfortunately we do not know the end of the story. Could the liver of a baboon really serve as a temporary surrogate for a human liver? Did it ever do so? What about the negative analogy? What sort of reaction would the baboon's liver cells have to the alien blood? What happened to the baboon? This is a modelling story that so far as I know has no known end.

Lower's dogs

By coincidence, two men, one living in the seventeenth century and one in the twentieth, who both used dogs as models and whose work led to major breakthroughs in medicine bore the same name. The first Richard Lower used dogs to perfect blood transfusion techniques, while the second Richard Lower used the same species of animals to perfect heart transplantation methods.

History, as presented in the media and so in popular belief, credits the beginning of heart transplantation techniques to Christiaan Barnard. Digging a little deeper we come across the most successful practitioner of this art, Norman Shumway. However, none of this would have been possible without the work of Richard Lower often but not always sharing the work with Shumway, using dogs as experimental apparatus—as pilot plants—to perfect the surgical techniques required to carry out transplant operations on human beings. Christiaan Barnard did forty-eight trial transplants with dogs before he undertook such an operation with a human being. He acknowledges that what he used 'was a technique built on that developed by Shumway and Lower, who had experimented on more than 300 dogs . . . With their findings joined to mine there was little point in continuing to further sacrifice of animals.'[4]

Richard Lower studied at the Medical School at Cornell but moved for his residency period to the University of Washington in the northwest of the United States. One autobiographic snippet suggests that the American northwest suited his love of the outdoors. Finding things not to his liking there he moved south to Stanford where he could qualify more quickly. Working in very primitive conditions in the training section of the hospital he eventually met up with Norman Shumway and forged a remarkable partnership with him.

Norman Shumway (1923–2006) studied medicine at Vanderbilt University, moving on to a doctorate at the University of Minnesota in 1956. He was appointed as a surgery instructor at Stanford University in 1958, where he remained for the rest of his career. He seems to have been a somewhat paradoxical character.

Reticent and wary of publicity, yet he was famous for his witty and jocular conversation, particularly during the course of long and demanding surgical procedures. The persistence with which he pursued his goal of successful human heart transplantation suggests a dogged streak.

Lower began his experimental programme as an assistant to Shumway working on the techniques for open heart surgery, including the possibility of cooling a living heart so as to stop its beating. Under this condition surgical repair would be greatly facilitated. The heart could then be warmed up and restarted with a fibrillator. Perhaps the heart could be removed completely from the patient's body for delicate surgical work and replaced when the repair work was done. This is the procedure known as 'auto-transplantation'.

Dogs were already in regular use at Stanford University Medical School for training surgeons. The workplace was an ill-equipped annex with cement sinks and a leaking roof. Undeterred by the dismal environment Lower and Shumway continued their use of dogs as experimental apparatus, pilot plants for the ultimate transfer of the techniques to the human case. Despite his skill in surgery Lower was unable to achieve successful autotransplantation. The reason was simple. To remove the heart from the dog the aorta and pulmonary vessels had to be cut in such a way that there was sufficient length on each side of the incision for sewing the heart back in. Lower simply could not manage to retain enough of the vessels on either side of the severing cut to meet this essential condition. However, it then occurred to him that in implanting a 'homologous heart'—one from another dog—the problem could be overcome. By cutting the aorta of the first dog close to the heart a substantial section of the aorta remained in the cavity of the chest. By cutting the aorta of the donor dog to leave a substantial length from the heart, enough 'tubing' was left to make successful suturing possible. The same held for other cardiac blood vessels. This technique proved to be entirely successful. The first operation transplanting a heart from one dog to another was carried out in 1959.

The results of the project were reported in the *Surgical Forum* of 1960.[5] In the discussion section of the paper the authors report that the dogs died within a few days and at best survived for as much as three weeks. The cause of death was found to be adverse immunological reactions rather than heart failure. 'If the immunological mechanisms of the host were prevented from destroying the graft, in all likelihood it would continue to function adequately for the normal life span of the animal.' Lower carried out other experiments, include a bizarre xenotransplant, inserting a human heart into a chimpanzee. A newspaper report described the result euphemistically: 'the heart functioned until Lower euthanized the chimp and terminated the experiment.'[6]

Lower eventually moved to Richmond, Virginia, where among his visitors was a South African surgeon, Christiaan Barnard. Barnard also visited Shumway whom he had known previously at the University of Minnesota. Back in Stanford it was not until 1967 that Shumway was ready to go ahead with confidence to a programme of human transplantations. During the preceding eight years he had continued his use of the bodies of living dogs as the essential apparatus for his transplantation experiments. It was a great surprise to him to learn that on December 23 his old acquaintance and recent visitor, Christiaan Barnard, had carried out a successful human heart transplantation operation on Louis Washkansky. The use of the necessary immunosuppressing drugs left open the possibility of fatal infection. Washkansky contracted pneumonia and died after only 18 days. After a few more operations Barnard abandoned his programme because the immunosuppressing drugs available at that time for preventing the rejection of the transplant also left the patient vulnerable to common and simple infections. Barnard too had practised his surgical techniques on the bodies of dogs. 'Scientific enquiry consisted in this: the use of knowledge to go on to further knowledge', he commented when discussing the need for the forty-eight canine heart transplants he had done before attempting the first human operation.

Shumway began his own programme of transplantations on 6 January 1968. While his patients survived the operation for increasing lengths of time, and given immunosuppressing drugs, continued on the path to recovery, the life expectancy of the recipients of healthy hearts was subject to the danger of infections with which the damped immune system could not cope. His first patient died after only 15 days. Unlike Barnard, Shumway continued, almost alone, to refine the procedure, including the post-operative suppression of the tissue rejection mechanism of the body of the recipient. Gradually the life expectancy of the patients improved. All this changed when a new drug, 'cyclosporine', was introduced. While suppressing the body's efforts to expel the alien heart tissue it allowed resistance to common infections to remain robust.

Mice are little people, too

Genetic experiments on human beings are, at least at the present time, outside the boundaries of the legally permissible. However, observing correlations between genetic make-up and behaviour is an increasingly popular speciality. So the question of whether there are close ties between the genetic make-up of a person and the character and personality dispositions and propensities they display in later life cannot be studied directly by manipulating the human genome. At best, one can study a person's genetic make-up and try to correlate that with personality and character traits and dispositions as they display them. The difficulty of this kind of research is compounded by disagreements among psychologists as to whether there are such attributes as personality *traits* and character *dispositions* that distinguish one person from another and that persist throughout the life course of an individual. Much recent research has led to an emphasis on the role of the situation and how it is perceived by human actors in accounting for what people do that would otherwise be taken to be indicative of personality traits. For example, the work of Phillip Zimbardo shows how much the situation in which people find themselves influences what they do and what they think it right to do.[7] People's dispositions change with the situation they find themselves in, the roles they are expected to play, and the people they encounter.

Even setting those doubts aside the question remains: how would one study the genetic basis of human dispositions and propensities? As often happens in the sciences the answer to such questions is 'make or find a model'. One would need an easily manipulable animal whose behavioural patterns are well known. Then by altering the genome of one or more such animals and studying the consequent patterns of behaviour one might have a basis for analogically transferring these results to people. Thus the animal, in many cases the humble mouse, serves as a model of the human being. However, sometimes nature provides the altered genome without any necessity for human manipulation. The logic of the use of the mouse model is exactly the same as if the mutant genome had been created by some laboratory technique:

what is the difference between the way mice with this genetic profile behave compared with those with a different but closely related profile?

Any inferences from genetic make-up to conclusions about the behaviour of an animal must take account of the nature of gene expression. Genes do not cause behaviour directly. Rather a gene, often in concert with other genes, is expressed in proteins that determine the character of some neural structure or biochemical pathway. The pattern of gene expression is part—but only part—of the mechanism that in suitable circumstances leads to characteristic kinds of behaviour.

MICE AND DEPRESSION

The way that research on mice is transferred by analogy to people is strikingly shown by newspaper reports of genetic research. According to Mark Henderson, *The Times* correspondent, 'A gene that may explain why some people are more likely to suffer depression has been discovered by scientists, paving the way for improved treatments of mental illness that strikes one in five people.... Though the mutation has so far been identified only in mice, the findings have excited mental health researchers.'[8]

Of course, considered scientifically this is reckless extrapolation, pushing the mouse model to or beyond its limits. Delving into the story we find two analogical links. The gene, about which more later, influences the production of serotonin in mice, a chemical that in human beings is known to be implicated in various aspects of human psychology. The analogical link suggests that the corresponding gene in human beings might also influence the production of serotonin. The second link is more tenuous. Is there any psychological condition in mice that matches depression in human beings? Only if that link is secure is there any virtue in the mouse model as an analogue of the nature and sources of human depression.

What role does serotonin play in human psychology? There seems to be a strong correlation between low levels of serotonin and tendency to depression. Indirect evidence comes from the effect of Prozac and similar drugs on people's sense of well being. These drugs prevent serotonin being reabsorbed and so effectively increase the amount available as a 'neurotransmitter'—a chemical involved in passing signals among cells in the brain. Neural signals move along nerve fibres as waves of potential difference between the inside and the outside of the fibre. Neurons meet at synapses. At the presynaptic end of the nerve cell or neuron the arrival of a signal stimulates the production of a neurotransmitter. The substance transfuses through to the synaptic end of a connected neuron, where it stimulates the beginning of a wave of potential difference that transmits a signal along the axon and via the synapse, to the dendrites of the next cell in the network, and so on. Serotonin is transported across the synapse by a transport protein. Unlike most neurotransmitters serotonin is produced and reabsorbed into the presynaptic terminal. This cycle can be affected by 'serotonin reuptake inhibitors' which target the serotonin transport protein. If the transport protein is inhibited then more serotonin will remain in the region of the synapse. In this way a deficit will be made good.

The relevant enzyme that limits the rate of serotonin uptake is tryptophan hydroxylase. Its expression in the brain derives from the gene Tph2 via a complex chain of events.

First let us look at the biological analogy. Marc Caron and his co-workers showed that a variation in the gene Tph2 affects the production of serotonin in mice. Mice with the variant gene produce very much less serotonin than normal mice. There are two forms of Tph2. The substitution of one for the other results in decreased serotonin levels in the brain. Caron and his team concluded that their data provide direct evidence for a fundamental role of Tph2 in brain serotonin synthesis.[9] What we don't know is whether this difference is displayed in character traits, in the behaviour of mice or people.

The first step to filling this knowledge gap is to select and experiment with a suitable model. Once again the mouse provides the step between biochemistry and human psychology. Luigi Cervo and his co-workers at the Institute for Pharmacological Research in Milan noted that less tryptophan hydroxylase is synthesized in the brains of mice with one of the variants of the Tph2 gene than the brains of those which have the other variant. They found that when mice are immobilized by a stressful event, such as being put through a 'forced swimming test', those with the variant that is expressed in less tryptophan hydroxylase remain immobile longer than normal mice. They took this to be a behavioural expression of 'mouse depression'. Once the mice were 'depressed', administering an antidepressant made no difference to how long they remained immobile. But if tryptophan was administered before the stressful event, thus bumping up the amount in the depression-prone mice, the antidepressant worked. This showed, they said, that the lack of success with the administration of the antidepressant, citalopram, to reduce immobility time in the mice with lower levels of tryptophan is attributable to the genetic variation.[10] But, what about people? We now have a second bridge to cross—from mouse to human being. This is the key modelling relation in the research as a whole.

Our doubts could be laid to rest if this research stimulated studies of the corresponding gene in people which showed that a similar change in Tph2 resulted in changes in the amounts of serotonin produced. Then we could delete mouse psychology from the story. There is at least some evidence for the psychological effect of different levels of serotonin in human beings. The research undertaken by Van Den Bogaert and co-workers from the University of Amsterdam, though limited in scope, looks promising in bridging the species gap and validating the mouse model. Shorn of the technicalities the research, which was conducted in Northern Sweden, showed that two groups of people, one with simple depressive tendencies and the other with a tendency to manic-depression (now often called 'bipolar disorder'), shared a deficiency in serotonin production. Comparing these people genetically with a control group who did not have either of these tendencies showed that Tph2 was associated with depressive disorders in sufferers of northern Swedish descent.[11] So far so good, but there is a long way to go before a truly general claim can be substantiated.

MICE AND AGGRESSION

Mice figure in another example of the modelling process which might illuminate how the psychology of human beings could be influenced by their genetic make-up. This research makes use of both the analogical links that framed the research into Tph2. Recently, a research programme of this kind has been undertaken by the husband and wife team of Avshalom Caspi and Terrie Moffatt with their co-workers at the Institute of Psychiatry in the University of London. In this case the focus was the gene labelled MAOA. The story starts with the discovery of a biochemical link between the MAOA gene and distinctive mouse behaviour.

This gene influences the amount of neurotransmitters at the synapses, including serotonin. If this gene is absent from the make-up of an organism, less serotonin is removed from the synapses, resulting in an excess of the neurotransmitter. It should follow that people in whom the gene is absent should be neurologically more active than those in whom the gene is present.[12] Unlike the Tph2 story, which turns on a deficit of serotonin, this story turns on an excess of serotonin and other neurotransmitters being found in the brains of mice.

Moreover, the researchers explicitly argued for those analogies between the behaviour of mice and people that would make the genetic research relevant. As we shall see there are many analogical links that criss-cross this research, making it possible for the results to serve as a model for the neurological background for certain human psychological traits. The richness of the analogy makes up for weaknesses in any one of the constituent links of the positive analogy.

The social psychological similarity between the behaviour of mice and certain human behavioural tendencies is pretty direct. The behaviour of male mice with the excess neurotransmitters was much more aggressive than that of biologically normal mice. For example, mice are territorial creatures and intruders into home territories were more quickly attacked by mice which lacked the MAOA gene than by those which possessed it. Furthermore, fights among male mammals usually result in long-term status hierarchies, in which challenges to and fighting for position is rare. Male mice lacking the gene, and presumably over-supplied with neurotransmitters, did not form stable hierarchies, and so fighting was much more common.

The mice formed an experimental apparatus interpreted according to the familiar methodology sketched out long ago by Francis Bacon. Mice which have the gene are less prone to fight than those who lack the gene. So, lacking the gene is among the conditions responsible for proneness to aggression, 'absence in proximity'.

So far we have been looking at the mouse model only. Is there a discernible link that would tie this mouse model to human beings? Clearly the ethological link will look somewhat different since it is unlikely that there will be much evidence of links between variants of the human genome and human territorial aggression. Defence of territory in the human case takes a huge variety of forms, from physical violence to court cases over the height of a Leylandi hedge!

People do not now and have not for a long time set up social hierarchies on the basis of the results of physical fights. Football hooligans might look like a counterexample to this generalization. However, not only is there no evidence that football hooligans are notably defective in the MAOA gene, but the 'fights' at such venues are largely ritualized.[13]

The direct evidence for the link between the presence or absence of the MAOA gene and human aggressive behaviour is pretty thin. The male members of a certain Dutch family who were particularly aggressive were shown to have a mutant form of MAOA. It is easy to see that this piece of evidence could be readily set aside. Did many other aggressive Dutchmen have variant MAOA genes? Was the pattern of gene expression such that it could be seen to be a necessary or maybe even a sufficient condition for proneness to aggression? How many other Dutch people with variant MAOA genes turned out to be especially aggressive? Still, one must start somewhere!

The research by Avshalom Caspi and Terri Moffat added a further element and a more convincing if more complex twist to the story. They found that only among men who had been abused as children did the absence of the MAOA gene and the consequential excess of neurotransmitters tie in with greater proneness to aggression. This adds an epigenetic rule to the whole picture. At least three stages are evident in the working through of the effects of the presence or absence of the MAOA gene. In the first stage, the expression of the gene via protein synthesis makes possible the activation of a biochemical mechanism, so in the absence of the gene that mechanism is not activated. In the second stage, the activation or otherwise of the mechanism affects balance and concentrations of neurotransmitters. In the third stage, the concentrations of the relevant neurotransmitters affects proneness to distinctive kinds of behaviour in well-defined circumstances.

According to Caspi and Moffat, 'Circumstantial evidence suggests the hypothesis that childhood maltreatment predisposes most strongly to adult violence among children whose MAOA is insufficient to constrain maltreatment-induced changes to neurotransmitter systems.' The second leg of the modelling relationship is set out as follows: 'Animal studies [for example, the mouse model] document that maltreatment stress... alters... neurotransmitter systems in ways that can persist into adulthood and can influence aggressive behaviours.' The model now serves as the basis for the inference that 'deficiency in MAOA activity may dispose an organism towards neural hyperactivity to threat.' The third leg links neural hyperactivity to the expression of excesses of neurotransmitters in violent behaviour.

Dramatic newspaper headlines such as 'Down in the Dumps May Be up to a Gene' are hardly well-grounded inferences from the scientific conclusions so far established by the work of Caron. Or the report that the 'warrior gene' had been discovered so we could now understand the impulse that took Menelaus and Achilles to Troy![14]

What does all this add up to? In the mouse studies these creatures have the status of a model or analogue of human beings. Manipulating or taking advantage of natural variations in the core objects of the model, namely the mice, a further parallel can be developed. The mouse genome as a model of the human genome becomes enriched as mouse behaviour is more or less successfully presented as a model of human behaviour. The scientific value of the research apropos of human psychopathology turns on the strengths of both modelling relations.

If we can't find an animal model a human body will do

One can generalize from the results of a study which began as a single case or idiographic experiment if one can convince oneself and others that the experimental setup and the result of manipulating it is typical of all other similar items in the domain of the generalization. One transforms the idiographic study into support for a conclusion according to the principles of the intensive design. By taking the specimens that one has already examined to be typical a large number of repetitions are avoided. Just a very few examples of a ring of fossilized filaments surrounding what seems to be a typical dinosaur skeleton is enough to accustom us to imagine some of them dressed in brilliantly coloured feathers.

Barry Marshall's use of his own body as an experimental apparatus illustrates the power of the well-chosen idiographic, one-off experiment to convince the sceptics, as it becomes transformed into an example of intensive design. In the case of the bacterium *Helicobacter pylori* and its role in gastric and duodenal ulcers there were plenty of sceptics. Marshall's model was not only a model for a human body, it *was* a human body. Logically the question of similarity and difference between model and subject could only have been concerned with how typical of human bodies in general was the Marshall body.

KOCH'S POSTULATES

To appreciate the accumulation of accidental discoveries that led to the moment at which Barry Marshall infected himself with the *Helicobacter* germ, we touch on a fundamental principle of medical science, Koch's postulates. These are the requirements that must be met by anything that is to count as the cause of a disease.

The first postulate requires that the suspected bacterium or virus is found in every case in which the disease develops and in no cases where it does not.

The second postulate requires that the suspected agent should be able to be grown in a culture independent of the disease process from which the agent has been isolated.

Clearly these postulates are good 'for the most part'.

Standing in the way of the immediate acceptance of any bacterial theory of peptic ulcers was the widespread belief that bacteria could not grow in the stomach because of the powerful acidity of the gastric secretions. If this belief were true no bacterial theory could be viable. Of course, if there were to be a bacterium adapted to flourishing in a highly acidic environment the way would be open for just such a theory. Today we are familiar with the extraordinary discoveries of the bacterial colonies around active undersea volcanic vents. The study of extremeophiles is a well-established branch of microbiology, which is discussed in some detail in Chapter 8.

The events which led up to Marshall's dramatic experiment go back to the nineteenth century when the Italian bacteriologist Giulio Bizozzero found 'spirelli', spiral-shaped micro-organisms, often deep in the internal organs of dogs. Born on 20 March 1846,

Bizozzero was a prodigy, publishing his first paper at 16 and being appointed Chief of Pathology at the University of Pavia when only 21. He published his discovery in 1892. Though well publicized at the time this particular observation later languished. The bacterium or one similar to it went on being discovered and rediscovered over the next 60 years. However, with no direct clinical evidence of its role as a pathogen, a source of disease, its existence was of no great moment.

THE PARTNERS

The collaboration between Barry Marshall and Robin Warren had something of the character contrasts that go to make an excellent research team, a psychological symbiosis we have already noted in several successful collaborations. Warren is a steady, methodical, and modest sort of person, while Marshall quotes with satisfaction a description of himself as 'brash'. Both were active and energetic explorers but each in his own way.

Born in Adelaide, South Australia, Warren graduated from the Medical School of Adelaide University in 1967. After a year studying pathology in Melbourne he started work as a pathologist in Perth, Western Australia. Studying a specimen from a gastric biopsy he noticed an unusual blue line around the edge. The microscope was just powerful enough to reveal 'numerous suspicious bacteria' adjacent to the damaged tissue. Bacteria reveal themselves under the microscope much more readily if they can be stained, differentiating them from the background of other material in which they are found. Trying out the Warthin-Starry silver stain Warren was able to see the bacteria very clearly. Further examples of gastric biopsy material showed these bacteria frequently in conjunction with gastric lesions as can be seen in Figure 12.

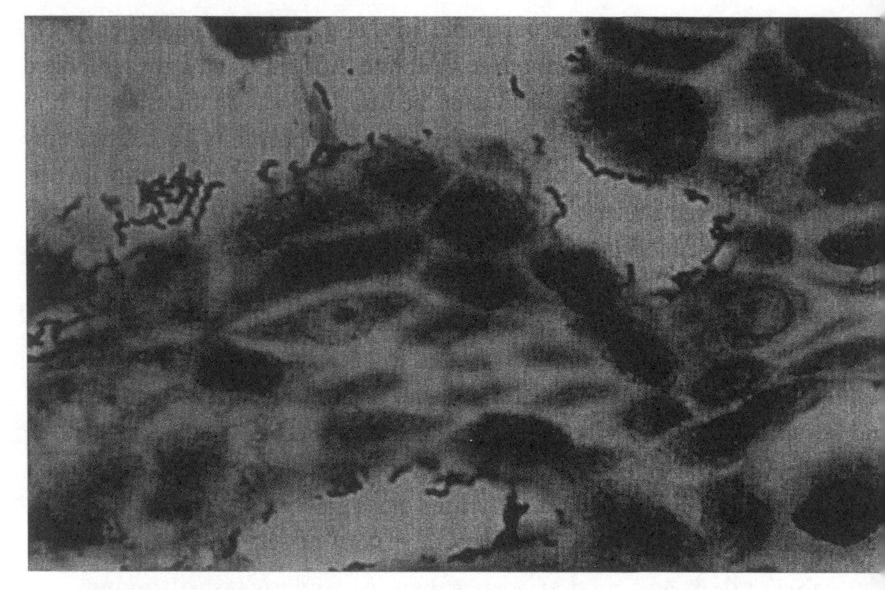

Figure 12. *Helicobacter pylori* visible in the stomach.

At this point the first hints of resistance to the obvious conclusion that gastric and duodenal ulcers were due to an infection by a bacterium appeared. As Warren says, his further researches were made difficult by the fact that the very idea of sending gastric samples especially for microbiological culture was considered ridiculous. However, the accident of the anomalous 'blue line' led straight to a reasonable case that Koch's first postulate had been fulfilled—these bacteria were present with the disease and absent when the stomach wall was healthy.

Barry Marshall was born in Kalgoorlie, Western Australia. He met Warren in 1981 shortly after he took up the post of Registrar at the Royal Perth Hospital, and there their collaboration began.

THE PARTNERSHIP

Koch's second postulate requires that the bacterium, seemingly implicated in the disease, be cultured independently of the disease process itself. Marshall's attempts to culture the *Helicobacter* organism independently had persistently failed. Then, on one occasion an assistant failed to turn off over the Easter break the machine in which the Petri dishes with the cultures had been incubated. When Marshall examined them, he had his positive result: the bacterium had been grown independently of its site in the gut of a patient. (This lucky accident is reminiscent of Pasteur's discovery of the attenuation of pathogenic agents. He had left a culture of chicken cholera in his laboratory for the whole of the summer break. Instead of throwing it out when he returned he decided to test whether it had any residual properties. To his surprise he found that chickens infected with material from the seemingly useless culture were immune to the disease.)

After some confusion over how to present their discoveries Warren and Marshall finally sent off two letters to the *Lancet* describing the results of their work. The publication was delayed and delayed because the editor could not find anyone to recommend acceptance of these letters. As Warren recounts in his Nobel acceptance speech, 'Barry and I completed our definitive paper and sent it to the *Lancet*. I believe the Editor wanted to publish the paper but the reviewers could not agree and held up publication for months: the findings were too new and strange.' The more easy-going Warren was unwilling to take drastic action to publish their results and set the record straight; but not so the 'pushy' Dr Marshall. It was this situation that led to his dramatic choice of his own body as the apparatus needed to test the

Helicobacter organism against both of Koch's postulates at once. Successful treatment of 'quite a few patients' gave him the confidence that he knew how to effect a cure for *Helicobacter* infection.

'Becoming increasingly frustrated with the negative response to my work I realized I had to have an animal model and decided to use myself', Marshall explained. By infecting himself with the bacterium and inducing the beginnings of a gastric ulcer to form, and then curing himself with the appropriate antibiotic, he would satisfy both postulates. The bacterium would be seen to be present with the disease, and, since the organism had come from another infected person, his own stomach would be a culture medium in which it would be shown to have grown independently of its origin. As further proof he would culture the bacteria obtained from biopsies of his own stomach while suffering from the self-inflicted disease.

But first he had to get independent confirmation that his 'gastric mucosa' were normal and free of infection. This required an endoscopy, a direct examination of the lining of the stomach by means of a tube inserted into the stomach with a viewing device on the end. Indeed, that morning in June 1984 he was perfectly healthy. On the same day he and his colleagues had obtained a culture of *H. pylori* from a middle-aged man with gastric symptoms. The organism was shown to be susceptible to the antibiotic metronidazole, and the patient was cured by this treatment. By this time a large amount of the *H. pylori* organism had been cultured. Using this preparation Barry Marshall 'swallowed about 30 ml of the solution of peptone broth just before midday on 12 June 1984'.[15]

A week later he woke in the early hours to vomit. This continued for the next three days. He was vomiting a clear liquid but it was not acidic. Ten days after swallowing the 'bug broth' he endured another endoscopy. The accompanying biopsy showed that the *Helicobacter* organism was indeed present. The report of the biopsy, shorn of medical terminology, described an inflamed and damaged stomach lining, and there too were the 'spiral bacteria'. Koch's first postulate was abundantly satisfied. The culture of the material from the biopsy showed that Koch's second postulate was also satisfied. Fortunately, two weeks after taking the 'bug broth' the organisms had disappeared under the influence of the antibiotic.

Barry Marshall recounts his wife's reactions to his self-inflicted illness: 'After putting up with me vomiting for a few more nights, moaning and groaning all night and having a breath like a sewer, Adrienne finally demanded that I immediately commence antibiotics or be evicted from the household to sleep under a bridge.'[16] The absence of excess acid with the disease lesions clearly showed that by Koch's first postulate, gastric disorders were not the result of excess stomach acid. That implied that acid-reducing therapies were not effective in curing the disease, however much they helped with the symptoms.

During this time Marshall and Warren's joint paper was published in the *Lancet*. After a celebratory dinner with their wives, Robin Warren was rung up at 5am by a persistent journalist from an American paper asking whether the spiral microbe was indeed a causal agent in gastric diseases. Perhaps a bit impatiently, Warren answered that indeed it was, and that 'Dr Marshall had proved it by drinking a brew of the bug, thus infecting himself.' The subsequent newspaper article led to worldwide publicity and in the end, after a long struggle against a sceptical medical establishment, a change in the way gastric diseases were treated was finally achieved.

Why the delay in the publication and acceptance of the results of Warren and Marshall's work? Was there a conspiracy among the drug companies to maintain their profitable line in acid-reducing medications? Almost certainly not—rather the delays were 'fired more by inertia and bad advice than by malice'. For a while the drug companies continued to fund trials of acid-reducing medications, but as patents ran out they turned to the development of antibiotics appropriate to the attack on *H. pylori*, the true villain of the piece. In 2005 Robin Warren and Barry Marshall shared the Nobel Prize for Medicine.

Commentary

The studies described in the first part of this chapter are samples of a vast amount of research built on the use of animals as models for human beings. There are several analogical links between the bodies of animals and human bodies. In the case of the use of dogs as models for research the analogy includes both anatomical and physiological similarities. They underlie the use of dogs in the preliminary studies that led up to successful human heart transplants. Another link matches the complex sequences that lead from a certain pattern of genes to certain kinds of abnormal animal behaviour to those which might be involved in human tendencies to psychiatric problems. The key link is through similarities in the biochemistry of the nervous system through which genetic endowment is ultimately expressed. However, the value of the studies of mice for our understanding of human life depends on analogies between behaviour patterns in animal and human contexts. The amount and prominence of newspaper coverage of current research indicates considerable public interest.

The spread of popular fascination with genetic psychology is well captured by these remarks from Jeffrey Eugenides.

> Back then [1960s and 70s] . . . everyone wanted to go unisex. The consensus was that personality was primarily determined by environment, each child a blank slate to be written. My own medical story [involving the danger for a hermaphrodite of being surgically 'reassigned'] was only a reflection of what was happening psychologically to everyone in those years. Women were becoming more like men, and men were becoming more like women. For a little while during the seventies it seemed that sex difference might pass away. But then another thing happened.
>
> It was called evolutionary biology. Under its sway, the sexes were separated again, men into hunters, women into gatherers. Nurture no longer formed us; nature did. Impulses from hominids dating from 20,000 BC were controlling us. And so today you get the usual simplifications. Why can't men communicate? Because they had to be quiet on the hunt. Why do women communicate so well? Because they had to call out for one another where the fruits and berries were, and so on. Men like blue, the colour of hunters' sky; women like pink the colour of the ripe fruit they must gather![17]

That an animal model need not be an animal and could even be the scientist himself is an insight given to few. Barry Marshall and Robert Warren knew the scientific virtues of the use of a model in which the stringent logical conditions expressed Koch's postulates could be fulfilled. Though Marshall would have used another means had one been available, his choice of his own body for this role displayed the animality of the human organism, as well as creating a model in which the analogy relation became identity.

Chapter 8
Modelling Worlds: Voles, Peas, Fruit Flies, Finches, and Deep Oceanic Worms

In the last chapter the models were narrowly focused on individual things and processes taken out of the world as a whole. In this chapter we will look at various models which attempt to represent the material basis of large-scale natural processes. The Oxford biologist John Clarke created two model mammalian worlds. In these miniature and enclosed worlds creatures lived out their lives in environments similar to those in which their feral cousins live. The great geneticist Theodosius Dobzhansky created insect models that represented the whole organic universe. Peter and Rosemary Grant and their assistants found and exploited a model bird world which represented the key process that had occurred again and again throughout the history of the biosphere, in a mere decade on a tiny island in the Galapagos Islands. Natural selection. Unknown to the fruit flies, the finches, and the voles, the inhabitants of these model worlds, the scientists, like Olympian gods, looked on from above, studied their activities, and recorded their vital statistics.

It might seem obvious that model worlds are to be read as analogues of the real world from which they are derived. However, in the cases to be described in this chapter each has grown in significance as the topic worlds have enlarged. What actual worlds are the model worlds models of? Is the Wytham Woods vole world that John Clarke constructed just a model of the real vole world, for example the grasslands of central Wales? Is it a model of the whole mammalian world? Could it even be an analogue of the animal world in general? The study of extremeophiles, living creatures surviving under conditions of extreme heat, pressure, or cold in contemporary environments, can be used to give some idea of the nature of the primitive organisms that existed in the early history of the Earth. If the choice of examples is right, the organisms that survive and even flourish under these conditions provide a model of what the inhabitants of Earth once were like.

Modelling the world of the vole[1]

My family once had a large black tomcat, called Nero. He soon learned that suitcases and other stuff at the bottom of the stairs meant that he would be in the care of the housesitter for the next few weeks. However, he also believed that giving ritual offerings to the 'top cats' in the house was an appropriate way to ingratiate himself with us. Soon a variety of small dead animal treats appeared. Usually they were mice, occasionally a shrew. Very rarely a different type of creature, something mouse-like but clearly not a mouse, would be there on the doormat. Eventually we found out that it was a vole. The village is close to open fields and small coppices, just the place for these animals to flourish. But we never saw them except as Nero presented them in the hope of changing our minds about going abroad. What sort of lives did these creatures live? What went on in the vole world?

Jakob Von Uexküll (1864–1944) was a distinguished Estonian biologist. Among his many contributions was the introduction of a very useful word for the region of the world inhabited by a particular species of animal, fish, or bird: its umwelt. Apart from their sacrificial symbolic role in the life world of the domestic cat we knew nothing of the umwelt of the vole. By creating a model 'vole world' John Clarke was able to throw a great deal of light on the umwelt of these creatures. The question he addressed in his extensive studies of his model world concerned the effects of crowding. How did the relative population density of a vole world affect the reproductive powers of voles, and so, indirectly via natural selection, the characteristics of successive populations?

John Clarke's interest in the natural world was kindled very early in his life on expeditions with his geologist father. At the University of Western Australia his fourth-year research project involved aspects of the anatomy of the wallaby, a small kangaroo, which he had come across on family holidays to the island of Rottnest (Rats' Nest) a few miles west of Fremantle, the port adjacent to Perth. The island was given this name by the Dutch explorer Wilhelm de Vlaming. When he landed on the island he mistook wallabies for large rats.

In 1947 John Clarke came to Oxford on a Rhodes Scholarship and spent his entire academic career there as a zoologist. The work on voles included detailed studies of their patterns of reproduction including the anatomical and physiological effects of stress on the reproductive organs themselves. Research on the biology of reproduction in the vole world led to public service for similar aspects of the biology of the human world. He became an Inspector for Human Fertilization and Embryology Authority, and later participated in an inspection of the Shanghai Institute for Planned Parenthood Research by a World Health Organization team.

The story of the suicidal migration of lemmings, relatives of the British field vole, is a highly coloured version of a feature of most natural communities—the cycle of population growth and decline.[2] Some years there are lots of wasps to plague us in the garden, in others hardly any. The biologist Dennis Chitty studied population cycles of field voles in Central Wales and showed that there was a four-year cycle as numbers rose and fell again. How was this cycle to be explained? If this was understood, perhaps generalizing from voles, population cycles among animals of similar species, such as lemmings, could be explained along the same lines.

Common sense suggests a very simple explanation. There is only so much food available for any species in its umwelt. According to Malthusian Law while populations increase by geometrical progression food supplies increase, if they do at all, only by an arithmetical progression. It follows inevitably that as the population grows so the food supply diminishes relative to population growth. The result is famine and an increasing death rate. The population declines until another cycle begins. Regular epidemics could also be a factor. The Plague returned again and again to Western Europe and the human population see-sawed in harmony. It has been suggested that the Black Death and the Plague were different diseases. Bubonic plague is spread by rats. Though the Black Death struck Iceland in medieval times there were no rats there until the nineteenth century. The fourteenth- and seventeenth-century epidemics had drastic effects on the population whether or not they were the same disease. There are many other possible influences that could bring about such populations cycles. The sizes of the populations of foxes and rabbits no doubt show cyclical changes. Perhaps weather conditions play a part. None of these seemed adequate to account for what Chitty found in his field studies. Perhaps the form of life of the voles themselves played a significant part. As the population increases so does the likelihood of fights breaking out over territorial boundaries, mating rights, and so on. Crowding leads to stress—a vague but fruitful concept—and stress could lead to lower fertility and shorter lifespans. If there are fewer voles to compete for the good things of life, fertility increases and lifespan grows. After a year or two with the population increasing all the time, we are back to the crowded conditions of the past—and so it goes on.

What sort of experiment could possibly test this idea? One could hardly extract a neat pair of factors, one to be manipulated as the independent variable while we track corresponding changes in the dependent variable. Boyle's experiments on the spring of the air were no guide, measuring the way changes in pressure went along with changes in the volume of an enclosed sample of gas. With the kind of intuitive sense of the way forward, the instinct that Michael Polanyi called the 'tacit dimension', John Clarke set about building a model vole world. In this world the aggressiveness of voles played into a complex pattern of demographic, behavioural, psychological, and physiological factors. He and his collaborators would dip into this world from time to time, extract an animal, and examine it. At the same time, peering in from above, they would watch the creatures go about their daily lives, all unaware that their world was an artefact.

To replicate the four-year population cycle two model worlds were needed. One would be crowded, representing the high point of the cycle, and the other would be thinly populated, representing the low point. Unlike the Grants' deliberate selection of an island as a model world and Dobzhansky's constructions of model insect worlds, Clarke's worlds came into being through a lucky chance. One day, walking with Dennis Chitty in Wytham Woods, Oxford University's Biological Reserve, they came across an abandoned and apparently forgotten swimming pool. Scouting about they came across a matching empty reservoir, also abandoned. By another stroke of luck the pool and the reservoir were exactly the same in surface area, 67 m^2. They had apparently been built during the Second World War at the behest of the headmaster of a school that had been evacuated to a nearby lodge. It immediately occurred to them that, secured from predators common in the Reserve, these structures, the empty pool and the reservoir, could be used to create habitats for experimental populations made up of animals drawn from Chitty's laboratory breeding colony. These were descendants of voles he had caught in Welsh grasslands. One could be the crowded vole world to be compared demographically with the other, set up as an under-populated one. Such a model might be used to demonstrate that intraspecific factors have an important role in the regulation of population sizes.

In a long tradition from Theodoric's model rainbow, to Newton's optical equipment and to the apparatus in Faraday's laboratory John Clarke set about converting his abandoned structures into model vole worlds, doing most of the work himself, like his distinguished predecessors. However, a model world is just that, a model of the world. In order to build a model one must make a shrewd judgement as to the balance of similarities and differences between the model and the reality which it models. If the differences are too great and the similarities too thin, the model may be too weak for any useful scientific purpose. But if the similarities are too great and the differences only minor then the model may be so choked with detail as to be rendered useless. There are no formulae for deciding the balance between similarities and differences—that depends on the intuitions, experience, and flair of the model maker.

So Clarke's first problem in transforming an old swimming pool and its reservoir into model worlds was to decide which aspects of the real vole world to leave out of his constructions. The possibility of ranging freely over a wide territory was eliminated. Each world was 67 m^2. More importantly Clarke eliminated predators from his model worlds—foxes, crows, weasels, etc.—by fixing wire netting over both the pool and the reservoir. To keep the pool dry Clarke hacked a drain hole in the wall of the pool. Much of the material needed for these jobs was begged or borrowed. Clarke named the two vole worlds 'The Swimming Pool' and the 'The Snake Pit'.[3]

The model worlds had to be furnished with material resembling the real vole world. Voles inhabit grasslands and create networks of tunnels through the vegetation. Locally harvested grass and bracken covering the floors provided a simulation of the grassland of the vole habitat. The animals soon created a system of tunnels for themselves within this cover. In addition a number of arrangements of bricks, enclosing a small space entered by a wooden tunnel, became the invariable nesting and social sites of the animals. Flap doors in the tunnels were made in such a way that at the time of each monthly census only inward traffic to these nest boxes was possible. This enabled Clarke reaching down from the human world, to count the animals systematically. A supply of whole oats, which voles devour avidly, was available at all times. Fresh cabbage was distributed around the cages twice a week, and there was a constant supply of water.

Studies in much stripped-down vole worlds had already shown that there was a marked propensity for aggression in both male and female voles. Clarke devised a system of glass tunnels in a smaller laboratory-based model of the grassland vole world. When sexually mature males encountered each other they fought fiercely, wrestling and biting. Even a whiff of the odour from the territory of another male set off incipient fighting moves. Females fought equally fiercely against intruders into their domains when pregnant or nurturing their young. In the model vole worlds, simple observation showed that the more crowded the world, the more aggressive encounters took place. There was also a decline in fertility, with fewer offspring being born relative to the population. In the sparsely populated world there was less aggression and greater fertility. At the same time in the laboratory Clarke investigated the ethology of voles cataloguing and analysing posture, gesture, and other signals displayed by voles in their aggressive interactions (see Figure 13). As in many species of animals and birds there is a clear distinction between the dominant and subordinate members of a group of non-specifics. When a dominant male vole is approached by a subordinate the dominant 'may make quick, "agitated", jerky runs', adopting a hunched attitude, gnashing of teeth and with hair erect, before eventually chasing and attacking the intruder. Sometimes severe wounds may be inflicted. However, retaliations do happen of course.[4]

The vole population established in the Swimming Pool was labelled P1 and that in the Snake Pit, P2. P1 comprised two males and one female, the unequal numbers due to the death of a female before the experiment really began. P2 was made up of six males and six females, four times the size of the P1 population.

Over the period of 'running the worlds' a weekly census was made of all the animals, which included the occurrence and size

Figure 13. Toilet in an aggressive situation. a), subordinate before approach of dominant. b), approach of dominant (right); squatting retaliation of subordinate. c), dominant retires, subordinate licks hands and brushes nose. d), dominant re-approaches, each vole bares teeth and squeals.

Figure 13. Vole postures.

of the litters. Both populations increased in size with the onset of breeding in the spring and its continuation until the autumn in the two breeding seasons during which they existed. P1 reached 23 animals and P2 58. It had been proposed to run the parallel worlds for at least two years, but after nineteen months fate intervened. Enterprising as ever, a weasel made its way through the netting into both worlds, killing all the members of one and part of the population of the other. It was time to pull the results together. Apart from this catastrophe, things had gone very satisfactorily.

Both males and females in the larger population had a shorter breeding season than those in the smaller group. Survival rate of infants and juveniles might be expected to be an important variable affecting population growth. Estimates of this factor were based on the number of offspring still alive after 15 days. Survival rates of infants and juveniles within each population were less than a corresponding figure for adults. However, between the populations there was no marked difference in the survival rate of any of the age classes. Given the pampered lifestyle in both worlds it was not surprising that there was no difference between the populations in survival rates for the young in the first phase of the experiment. Yet the statistics changed with time in P2, the crowded world. In the second breeding season the probability of an infant surviving was 0.46, as compared with the figure of 0.71 for the first breeding season. Furthermore the fertility of the females of P2 was less than that of the P1 females, the sparsely populated world. The increase in population of the crowded world was slowing down.

Summing up the results Clarke remarks that 'although both populations of voles expanded from their initial size, the growth rate of P1 appeared to be greater than that of P2 . . . something tending to limit the growth of the populations was acting more severely on P2 than on P1.' The physical conditions of the two populations were more or less the same. But P2 was larger than P1, so the limiting condition had something to do with population size relative to the fixed environment. Remember that the derelict pool and the reservoir were equal in area.

The key to the mystery lay in the observations of the way voles behaved. P2 animals fought more often and more savagely than those of P1. Perhaps fighting was the restraining factor. Yet, adults had good survival rates. Whatever effect fighting was having on the breeding powers of the voles must have been occurring before maturity. That fertility was affected was shown by the fact that second-generation females brought up in the crowded world of P2 were less fertile than those of the initial population brought up elsewhere.[5] It was clear that the animals in P2 had more aggressive encounters than those in P1. Perhaps this difference in behaviour could be at least part of the explanation for the demographic differences between the two model worlds. Fighting is manifested in scarring. Animals in P2 had more scars than those in P1. Many animals were scarred on their hind quarters, while only some were scarred on the head. This difference in distribution was surely the result of retaliation by those attacked wheeling round to face up to the dominant animals who had been pursuing them.

This was all very well as a contribution to ethology, the scientific study of animal behaviour as developed by Konrad Lorenz and Nikko Tinbergen, but how could this be linked to stress? Was there a mechanism of sorts that linked the observations into a pattern? The best candidate for such a mechanism would surely be glandular. The third phase of the research embraced the objective measurement of stress in voles brought about by aggressive encounters.

'Stress' is one of those words used to comprehend a broad domain of uses, linked, as Wittgenstein put it, by 'family resemblances'. It can have psychological, physiological, and even engineering applications. Inspired by Hans Selye's work on the aetiology of stress as a disease, Clarke devised a method of gauging the stress in voles which had been subject to aggression.

It had already been established that stress provokes the activity of the adrenal glands, bringing about dramatic increases in the secretion of steroid hormones from the adrenal glands and the thymus. By comparing the weights of the adrenal glands in stressed and unstressed animals a link between the behavioural components of aggression and its physiological consequences could be demonstrated, with a quantifiable variable, the relative weights of the glands. The relative weights of the thymus gland in the two populations could also be used as evidence. These procedures became part of the research programme.

Thus the third phase required a direct comparison between the relative weights of the adrenal and thymus glands of stressed and unstressed voles. The experiment required another stripped-down vole world, with tunnels and 'home areas'. Let us call the defenders of a territory the 'home' voles and the strangers the 'visitors'. The home voles attack the visitors, sometimes savagely. What effect did this have on the adrenal and thymus glands of visitors, the elements of the endocrine system most likely to be affected by fighting? Another group of animals were kept isolated from any fighting as matched controls. Did the adrenals and thymus of the animals which had been attacked, and so presumed to have been stressed, differ in weight from those unstressed animals which had not? Of course, a single incident could hardly be expected to have much effect, so the encounters were allowed to recur over a matter of hours and even days.

The adrenal glands of the animals subject to aggression were significantly larger than those which were not, and the thymuses were significantly smaller. The result demonstrated that fighting is stressful. The greater the duration of fighting, the greater the stress on at least some of the animals, the subordinates. Behavioural stress modulates into the neurophysiological events affecting the adrenal glands. The enlargement of these glands in the experimental animals might be read back into the vole worlds of the Wytham Woods; more crowded conditions bring about more encounters and so more fighting. Only one link is still missing—how does enlargement of the adrenals affect the fertility of females and the survival rates of infant voles?

Darwinian theory requires that the dynamics of species change be looked for in the relative fertility and survival rates of variant members of a species. In later studies Clarke and his collaborators undertook a study to track such changes. However, for this purpose they constructed other model vole worlds with less overall similarity to the real vole worlds from which all such models ultimately come. By altering the amounts of light and darkness in which a vole population lived and bred they looked for the emergence of stable breeding lines.

John Clarke's model worlds, representing the high and low points of population density in the life cycle of voles, are not a simple analogue of an experimental apparatus. The vole worlds did not test Chitty's hypothesis, that features of the lifestyle of vole worlds are responsible for the population cycle, directly. The situation is much more interesting. Taken together, the Swimming Pool and the Pit constitute a time-evolving model vole world for which Chitty's hypothesis is a defining principle. These worlds are such that aggressive relations between individual animals are constitutive of the life forms lived in those worlds.

The successful mimicry of the population cycles in real vole worlds supports the idea that they too are constituted by fighting males and squabbling females.

Modelling the world of the fruit fly

Pavlov's dogs were pieces of apparatus, as much as the arrangements of glassware in the chemistry laboratory or circuits and resistances in the physics laboratory. In the case of the fruit fly, whole colonies with quite large populations were the living apparatus. *Drosophila* colonies were used for the study of genetics, first the patterns of heritable characteristics and later the bearers of those characteristics.

The Darwinian theory of evolution by natural selection depended on the existence of variations in the populations of living species generation by generation. Mendelian genetics was focused on the repetition and constancy of the characteristics of living species. One scientist above all showed, with the help of the use of drosophila as an indispensable item in the instrumentarium of biology, that a synthesis of the two root theories of biology was possible, indeed inevitable. This scientist was Theodosius Grigorievich Dobzhansky.

THE DISCOVERY OF THE LIVING CELL

Genetics is the science that undergirds our observations of the patterns of heredity. A crucial initial step necessary for the insights and research programmes that have led to our present understanding of the processes by which plant and animal characteristics are passed down was the realization that all, or almost all, organic beings were cellular. This was not realized until the seventeenth century when, in 1665, Robert Hooke (1635–1703)

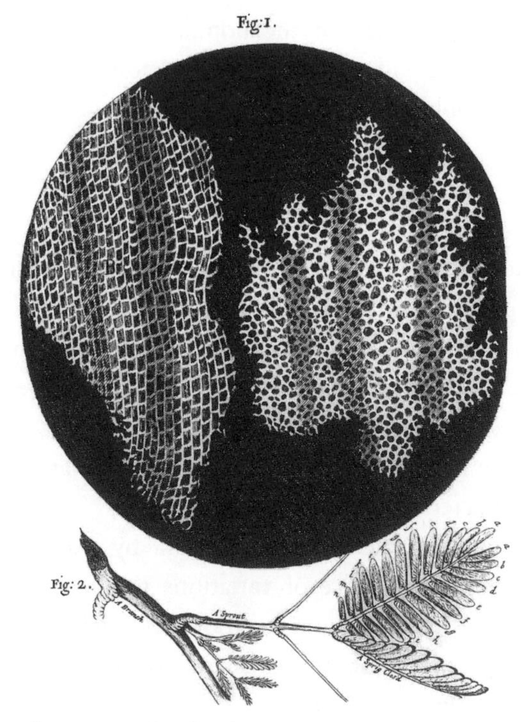

Figure 14. Cells in a sample of cork.

published a remarkable book, the *Micrographia*. Hooke himself drew the illustrations as he peered through his microscope at a hitherto almost unknown realm of minute beings. Among many discoveries, he saw for the first time that animals and plants were composed of myriads of cells, like the bricks that make up a wall. One of his drawings is shown in Figure 14.

An animal or plant grows from an embryo to an adult. But how? It seems obvious now but it was only with difficulty that growth was realized to be primarily a matter of the multiplication of cells. Where there had been one, there were now two, four, eight, and so on. So whatever determined the forms and qualities of adult organisms must somehow be present in the very cells which multiplied into the fully grown plant or animal. The fundamental location of inheritance must, it seems, be found in the multiplication of cells.

GREGOR MENDEL AND THE PEAS

As 'every schoolchild knows', the Abbé Mendel discovered the most basic laws of the patterns of inherited characteristics. However, as we shall see, his choice of peas as his experimental organisms guaranteed a very leisurely pace for his research programme into the patterns of heritable characteristics generation by generation.

Jacob Mendel was born in 1822 on a farm near Heizendorf in the then province of Austrian Silesia. His family supported him at the nearest boarding school at some sacrifice to themselves. The curriculum was broad, though old fashioned. In 1843 he entered the Augustinian order, adopting the name 'Gregor', and for fifteen productive years he taught at the Modern School at Brün (Brno). However, he did not manage to pass the qualifying examinations in science, so he remained a 'substitute teacher' for the whole of his pedagogical career.

His interest in plant breeding led to his being given a garden plot for his exclusive use. There the famous experiments on heredity were undertaken. His first important result was the demonstration, by transplanting a wild species to the garden, that plant forms did not depend on soil type and quality. His first efforts to study inheritance by experimenting on the colours of gillyflowers led nowhere. Eventually he chose peas. These had some advantages. Large crops were possible, there were many varieties, and they were easily artificially fertilized, so the ancestry of successive generations could be easily controlled. In 1854 he began the study. Pea plants display a number of fairly well marked bipolar qualities, such as tallness or shortness. The peas themselves could be green or yellow, smooth or wrinkled, with smooth or ridged pods, and so on. Mendel recognized eight such pairs in all.

The breakthrough came in crossing tall and short varieties. The next or hybrid generation all turned out to be tall. Surely, a mixture of shortness and tallness should have led to middle-sized plants. Breeding from this generation Mendel arrived at his famous ratio, 3 tall:1 short. He had discovered the phenomenon of dominance and recessiveness. The genetic factor expressed in the adult plant as the attribute 'tall' was dominant over that which would be expressed as the attribute 'short'.

In the modest paper he published in an obscure journal reporting these discoveries he gave the figures 787 tall to 277 short. Passing to a count of the peas themselves he got 6,022 green to 2,001 yellow. Similar results came from the count of the other binary qualities.

Algebraically we can picture the process something like this:

Let *A* be the dominant genetic factor and *B* the recessive. Then when an AA and a BB plant are crossed the first hybrid will consist of plants with an *AB* genetic make-up. Since *A* is dominant over *B* all the plants of this generation will display the attribute expressing the *A* factor, that is, all will be tall. However, in the next generation the segregation of *A*'s and *B*'s in the reproductive process will lead to the distribution

AA, AB, AB, BB

Since *A* is dominant over *B*, we get the familiar Mendelian ratio of 3 tall plants to every 1 short one.

Mendel's results are extraordinarily neat. Some have doubted that the schoolmaster actually got these ratios. Replications of the experiments have given results more divergent than Mendel's neat integral ratios. Recently a brilliant replication by Root-Bernstein has rescued Mendel's reputation.[6] Would the teacher have sat down and counted thousands of peas himself? It is much more likely that the boys in the class were given this tedious task. Suppose that they had three bowls, one for definitely green, one for definitely yellow and one for intermediate shades. Root-Bernstein showed that if the peas in the third bowl were randomly distributed between the yellow and the green bowls at the end of the count Mendel's results were replicated.

Mendel eventually took over the running of the Augustinian Abbey of St Thomas in Brün (Brno) where he remained for the rest of his life. He died on 6 January 1884, practically unknown in the world of biological science to which he had made such a fundamental contribution. He was succeeded by a man jealous even of Mendel's modest reputation. The next abbot had all Mendel's records destroyed.

Just as in other examples discussed in this book, Mendel's researches can be read simply as tests of a hypothesis, that heritable characteristics are carried by discrete natural units. Each observable character of a plant is the product of the presence of one type of genetic unit. However, Mendel's garden can also be thought of as a representation or model of the whole plant world. Later research in genetics tended to favour the model world interpretation of the living apparatus.

THE TURN TO *DROSOPHILA*

Modern genetics starts from 1900, when Mendel's results were rediscovered. But why *Drosophila*? The successive generations of peas appear at yearly intervals. The Abbot was in no hurry. However, others were not so patient. The various species of fruit flies breed at a much more satisfactorily rapid rate so that generations succeeded one another every 2 weeks. From egg to maturity is a matter of about 8 days. The laboratory apparatus that Mendel had constructed out of a garden plot and a few thousand pea plants sufficed for the original studies, yielding the first glimpse of the mathematical laws of heredity. A new apparatus would be needed to carry the work further.

Some 60,000 papers have been published to date on the genetics of the common fruit fly, *Drosophila melanogaster*. This rapid breeding species first attracted the attention of biologists in the United States. The first recorded use of a population of fruit flies to study inheritance patterns was by William Ernest Castle (1867–1945) in 1901. He used laboratory-breeding populations, following the distribution of characteristics generation by generation, much as Mendel had done with peas in the Abbey garden.

Who was Castle? History reveals a somewhat endearing character. He seems to have routinely taken the losing side in the scientific controversies of the day, yet the generosity and good temper with which he acknowledged his errors and embraced the better established alternative is exemplary.

After studying the Classics and later teaching Latin at several local institutions, he entered Harvard in 1890 to study biology. He took his PhD in 1895 working on a topic in embryology. In a precursor of the recent Nobel Prize-winning research on the development of the nematode worm by Sidney Brenner, Robert Horvitz, and John Sulston, Castle followed the fate of every cell as it divided and re-divided in the early stages of embryological development. In the course of this work, he realized that in this hermaphrodite species he had the first example of self-sterility in animals. Though each individual produced both male and female gametes, the eggs of one individual could be fertilized only by the sperm of another.

At first hostile to some of the underlying principles of Mendelian genetics, he eventually adopted them himself. To explore the effects of inbreeding and outbreeding over many generations he set up an experimental programme using a colony of *D. melanogaster*.

Though we must give Castle the credit for the choice of this insect as a piece of apparatus from the living instrumentarium, his main interest lay in experimental tests of the Lamarckian theory of the inheritance of acquired characteristics. Using the colour variants of the guinea pig, he set about proving that the genetic material within the germ cells and responsible for the colours of adults was unaffected by their bodily environment. With the help of a colleague's Pavlov-like surgical skills, he transplanted the ovaries of a black-haired guinea pig into the body of a white animal from which the ovaries had been removed. Luck went his way and the transplant took. All the offspring of the white female with the ovaries of the black animal were black. After his retirement he began a Mendelian analysis of the colours of the coats of horses. He died in 1945.

Why did Castle choose *Drosophila*? The species of this genus of fruit fly exhibit a number of adult characteristics that have the same logical pattern as Mendel's smooth and wrinkled, yellow and green peas. Though all four cross combinations are possible, the dominance/recessive distinction is needed to explain the patterns that one actually observes in adult organisms, generation by generation.

The transformation of *Drosophila* from a wild creature into a laboratory apparatus was completed by Thomas Hunt Morgan, in his laboratory in Columbia University in New York. He was born in Lexington, Kentucky, in 1866. Like many famous biologists he grew up in a more or less rural environment, in the hills of Kentucky and in the countryside near his mother's relatives in Maryland. He began his scientific career with detailed phylogenetic studies based on comparative descriptive embryology. The idea is simple: the history of the evolution of a species (phylogeny) is recapitulated in the stages of development of embryos of that species in the here and now (ontogeny). This has proved to be a useful half-truth. By 1895 he had abandoned the analysis of that which nature had provided and began intensive experimental studies of inheritance. Though he eventually won the Nobel Prize for his demonstration of the role of chromosomes in heredity, he first rejected the idea.

Despite its notorious squalor, his laboratory was a wonderful place to work, and attracted young men of talent. Personal reminiscences of his students and collaborators display Morgan as both dogmatic and predatory. Though he could hardly be accused of plagiarism, he made very free use of the ideas of his assistants, in a spirit of collective enterprise that was resented by some. It should be said on his behalf that he was happy for his own bright ideas to be taken up by his collaborators. Like Castle, he was often on the wrong side in a scientific controversy. The most important of these mistakes was his long-standing and vociferous opposition to the Darwinian mechanism of evolution by natural selection. He was also strongly critical of the idea that an accessory chromosome determined sex. Eventually he had to concede that females had two X chromosomes, while males had an X and Y.

Morgan's key idea was that the chromosomes must at least contain the genetic material. Though he was reluctant to admit it, the evidence of the work of his own assistants showed pretty clearly that genes must be parts of chromosomes, and what is more with specific locations on the chromosome. Sturtevant, one of his assistants, even as an undergraduate worked out a neat way of determining the distance genes must be apart on the chromosome to be consistent with the observational evidence of seeming genetic anomalies.

Once convinced of the role of the chromosomes Morgan threw himself into the research programme, choosing a colony of *D. melanogaster* as the living instrument of his researches. He made a startling discovery. A mutation of eye colour, the rare white rather than common red, turned out to be linked to the sex of the adult insect. Only females displayed the mutation. Here was a mechanism of variation different from the Mendelian distinction between dominant and recessive genes. Morgan quickly made the connection to the inheritance pattern of haemophilia and the other major sex-linked human characteristics, such as colour blindness. However, where were the variations on which natural selection could work?

In 1928 he was invited by Cal Tech to set up a Life Sciences Department. He moved his entire laboratory and his team to Pasadena. In that team was the man destined to solve the problem of the origin of the variations upon which Darwinian selection worked, Theodosius Dobzhansky. Morgan died in Pasadena in 1945, after a short illness.

Studies by Americans from the time of Castle have almost all been laboratory-based using populations in artificial environments. The same has been true of the great majority of animal studies, particularly work on primates. Harry F. Harlow made his name with experiments, described in Chapter 6, on patterns of mothering among monkeys and apes confined under strict laboratory conditions which privileged controlled experimentation over observation.

The story of Russian research into evolutionary genetics with *D. melanogaster* displays a very different nation-state/cultural tradition. Russians worked in a well-established natural history tradition, which we can see as part of a pan-European style in biological research. They studied natural populations of *D. melanogaster*. This work culminated in and around the years 1926–7. It was led by S. S. Chetverikov (1890–1959), who eventually fell victim to the machinations of Trofim Lysenko (see Chapter 9). In all probability the idea of using *D. melanogaster* as the research object had been suggested by the earlier American studies. However, in using natural populations Russian biologists broke not only with an experimental technique but also with a fundamental feature of the American idea of science. While American scientists favoured controlled experimentation Russian biologists were in favour of giving priority to disciplined observation.

Chetverikov's group took for granted that the Darwinian thesis of evolution by natural selection was the correct account of the origin of species. According to Chetverikov, it needed a basis in genetics. However, there were other views held by respectable scientists in the 1920s. In particular, the man who had been Dobzhansky's teacher, Filipchenko, was ready to admit other factors into the process by which the variants acted upon by natural selection were produced. Did the selection environment itself have some influence on the possible characteristics displayed by a species of adult creatures, be they plants or animals? In the 1920s these possibilities were still open. Theodosius Dobzhansky grew up in a milieu in which the mechanism of evolution was the prime topic of interest. Unlike the Americans with whom he later came to work, Russian biologists did not dismiss hypotheses concerning the mechanism of evolution as merely speculative; though they worked within the general theory of evolution by natural selection.

The possibility of the inheritance of acquired characteristics had been a vexing problem that had bedevilled the study of heredity for at least a century. Could characteristics that had been acquired by one generation of animals or plants be passed on to the next by inheritance? The idea that the environment can cause changes in the genetic material, thus enabling the inheritance of acquired characteristics, is associated with the French biologist Jean Baptiste Monet de Lamarck (1744–1829).

Lamarck had expressed this idea in a rigorous scientific form. He proposed a law of use and disuse summing up the way he believed the environment could affect an organism. If a bodily organ is constantly used for a certain environmentally imposed task, it will tend to develop to facilitate that use. An organ or body part which is disused will shrink and atrophy. The influential historian of science Charles Singer expresses a common view of Lamarck's work in his judgement that 'many of [Lamarck's] views were so fanciful that he was lightly esteemed by most of his contemporaries.'[7] However, they were not so lightly esteemed as not to be thought worthy of scientific tests. For example, in 1907, Fernandus Payne began a major research programme into Lamarckianism in T. H. Morgan's laboratory in Columbia University. Fernandus Payne (1881–1977) picked up the idea of using the genetic material of a population of *D. melanogaster* as the experimental focus, and the colony of insects as the apparatus. It is important to realize that these were not natural populations living under open conditions. They were confined to an artificial environment in the laboratory.

Again, the attempt to prove or disprove Lamarck's theses was strictly and perhaps necessarily laboratory based. Payne kept 40 generations of the insects in the dark. If Lamarck's thesis that the environment directly affected the genetic material had been correct, surely in that span of generational 'time' the eyes of later generations would have changed or even atrophied through disuse. Payne found no change at all. Compare this finding with the discovery of blind white fish in the subterranean aquifers that have been used for thousands of years as irrigation channels in Iran.[8] Surely, they lost their visual powers because, in some sense of that word, they had lived in the dark for countless generations. From a Darwinian standpoint there must be a trade-off between the cost of maintaining a useless organ and the adaptiveness of channelling that energy elsewhere. Perhaps this loss is a side effect of other adaptations to subterranean life.

THEODOSIUS GRIGORIEVICH DOBZHANSKY

Theodosius Dobzhansky was born in 1900 in the small town of Nemirov, near Kiev, in the Ukraine. He acquired the unusual name of 'Theodosius' because his then childless parents had visited the shrine of St Theodosius at Chernigov, vowing to name the child they hoped for after the saint. Dobzhansky's father was a mathematics teacher in the local school. As a schoolboy, both at home and on trips to the Caucasus Theodosius collected ladybirds, vowing to make a complete collection for each place.

He began graduate studies in the laboratory of Yuri Filipchencko (1882–1930), who nominated him for a Rockefeller award to spend the year 1927 in the laboratory of T. H. Morgan at Columbia University in New York. Accustomed to higher standards of laboratory hygiene and to a wide-ranging scientific discourse, Dobzhansky was disgusted by the filthy laboratory and dismayed by the narrow positivism of his colleagues. Morgan's passionate hatred of religion had drawn him reluctantly to the Darwinian theory of evolution though he was suspicious of its 'speculative' character. Evolution was the only wider topic which he and Dobzhansky could discuss. When Morgan moved his team to Cal Tech in Pasadena, Dobzhansky fell in love with the countryside of the Western United States and continued to base his work on the genetics of *D. melanogaster* but in the Russian style, studying natural populations in the rural areas he enjoyed. Yet, his synthesis of Mendelian genetics and the Darwinian theory of natural selection required intensive laboratory studies of the minute details of the mechanism of inheritance. In this phase of his work, he transformed the *D. melanogaster* colony into a new kind of apparatus from that which had been constructed by Morgan and Sturtevant.

It was thanks to a freak riding accident resulting in an enforced spell in hospital that his great book *Genetics and Evolutionary Biology* finally got underway, eventually being published in 1937.

Dobzhansky's project had little to do with the work being undertaken by Morgan and his group, except in so far as Dobzhansky was clear that genetics must be the key to understanding the mechanism of variation upon which the Darwinian theory ultimately depended. He brought this project with him from the Russian tradition in biology. Did evolution require both micro-selection within a species and a form of macroevolution in competition between species? Alternatively, would microevolution be sufficient? If it were enough, then the close study of the way mutations appeared in a single species would be the experimental key to the whole problem. Here, in the *D. melanogaster* colonies of Morgan's laboratory was a piece of apparatus perfectly suited to explore the processes and promises of microevolution. It might even prove so powerful as to eliminate the need for any other hypothesis.[9]

Dobzhansky's initial predicament at the moment when he joined the Morgan laboratory was having to cope both with a laboratory culture hostile to methods involving extensive field studies and with a certain hostility to evolutionary biology. The prevailing attitude was particularly distressing for Dobzhansky, for whom there was no necessary dichotomy between field and laboratory studies on the one hand, or between evolutionary theory and the rest of biology on the other. Nevertheless, Dobzhansky was able to take advantage of what the Morgan group had most to offer in experimental techniques and methods of research with *D. melanogaster*.

The contrast I wish to draw is not to be read as expressing a judgement one way or the other on the virtue of the laboratory method vis-à-vis fieldwork. The contrast between the European methodology of Dobzhansky and the American methodology of Morgan's assistant Sturtevant goes very deep into the basic presuppositions of contrasting cultural styles. The American style is invasive and active; the European observational and contemplative. The distinction of style is as marked in comparative medical practices as it is in biology. Each, of course, has its part to play.

The same insect species figures as the experimental apparatus in both traditions. In the laboratory, the insects are manipulated. In the wild, they are observed. As we have seen an apparatus is a model of whatever it is used to study. The breeding and genetic patterns of *D. melanogaster* and *Drosophila pseudoobscura* are meant to be models for inheritance in general. However, the distinctive styles of work of American and Russian biologists allows us to offer a more precise answer to the question of what it is in the real world the colonies of *D. melanogaster* are modelling. It seems that Russian studies are investigations of natural selection and genetics in nature, in the complex and ever-changing environment of the outside world. American studies, on the other hand, are investigations of domestic selection, of how stock-breeders and nurserymen refine and revise the animal and plant species that they have drawn out of nature into the simple and stable environments of laboratories or farms and gardens. *Drosophila*, as raw material, was turned into two different items of laboratory equipment by Morgan's assistants and by Dobzhansky.

The move to the laboratory to Pasadena unleashed some deep conflicts in Dobzhansky's mind as to whether he should stay or return to the USSR. Having decided to stay, he had troublesome run-ins with the US immigration service and the FBI. In the 1930s, the time of the Great Depression, the political Left was under suspicion, with Russians the obvious exemplifications of it.

To study the genetics of populations Dobzhansky had to develop a rapid method of analysis. He studied variations in the morphology of the testes with disposable microscopic preparations. This was in sharp contrast to the painstaking work of the Americans in the team. Here is how Kohler sums up the contrasting laboratory equipment that *Drosophila* yielded:

> It was a conflict not of good and bad practice but of different practices, new and old. Sturtevant [long in the laboratory] tried to domesticate *Drosophila pseudoobscura*, extending traditional laboratory practices and standards to a wild creature, controlling and confining its variability. Dobzhansky brought a wild creature into the laboratory, multiplying and exploiting its variability and transforming traditional laboratory practice. Dobzhansky and Sturtevant made *Drosophila melanogaster* into two different kinds of scientific instruments, two different modes of scientific practice, embodying different concepts of the relation between the laboratory and the field.[10]

Sturtevant studied the genetics of these insects as a topic in itself. The colony represented nothing outside itself, except in so far as it exemplified a concrete instantiation of the abstract Mendelian laws. Dobzhansky treated the laboratory colony as a model of the life of these creatures in the wild. Not only that, but as a model of the whole of living nature. Dobzhansky's target was nothing less than an understanding of the evolution of species, completing the work that Darwin had begun.

The difference between the two ways *Drosophila* colonies were transformed into apparatus can be seen in the way the experimental work was done. Dobzhansky developed a rapid and approximate method of following Mendelian patterns in successive generations. Precision was sacrificed to breadth of coverage. Dobzhansky's genius as a scientist consisted in part in his almost perfect judgement as to how the balance between precision and fruitfulness was to be struck in each of his major study programmes.

By 1940 Dobzhansky had become convinced that no mechanisms other than those described in microevolutionary theory are required to account for the whole of organic evolution in nature itself. The laboratory colonies as he managed them represented the whole of nature. This realization coincided with the opportunity to return to Columbia as a full professor with his own laboratory.

However, Dobzhansky, like many others, had become enamoured of the American West, and visited there every summer to continue his field studies of the insect. As his fame grew he was able to visit very widely, and he made an especially strong and productive linkage with Brazil. On his retirement from Columbia, he took up an offer from the University of California at Davis to move back west where he spent the rest of his life. He died in

1975. His ashes were scattered near the summer camp at Mather, California.

It was here that he and his family and generations of students and colleagues had combined wilderness holidays with intensive biological research in the real world. His daughter recounts how they released mutants marked with conspicuous red eyes in the middle of a cross-shaped pattern of traps consisting of small cups of fermenting mashed bananas on wire stands. The later recapture of samples of the wild population allowed the fate of deliberately introduced mutations to be followed. Dobzhansky combined the laboratory and the wilderness in a productive synthesis that finally cracked the problem of the mechanisms of natural selection.

Modelling natural selection in the real biosphere

Imagine if one could compress the surface of the Earth into less than a couple of square kilometres, might one not be able to watch natural selection in action, a process that Darwin could only imagine? This is one way of looking at the achievement of Rosemary and Peter Grant of Princeton University. They made evolution by natural selection visible. They showed how swiftly natural selection can alter biological traits such as the size and shapes of beaks of birds.[11]

To perform this miracle they had to find a living laboratory. They needed a huge apparatus in which they could follow the outcome of processes that were to serve as models for processes on a much grander scale in space and time. The island of Daphne Major in the Galapagos archipelago, the group of islands made famous by Darwin, became their working model.

What of the people who could envisage and actually carry through this amazing project? Fair weather or foul, months at a stretch, catching and weighing birds, counting seeds, and coming back year after year to go on with these onerous tasks, character must play a great part. Living conditions on Daphne Major were and indeed still are primitive—a single tent sufficed for living accommodation. All their food was brought in from Ecuador in freeze-dried form to avoid the risk of introducing seeds or insects not native to the island. After a degree in zoology from Cambridge, Peter Grant took his doctorate at the University of British Columbia for studies identifying trends in the sizes of certain island bird species on the Tres Marias Islands in Mexico. It was at British Columbia that he met Rosemary Matchett, also born in England and a research associate there at the time. Their groundbreaking study of natural selection on one of the Galapagos Islands began some years later.

Beginning in 1972 every year for many years the Grants returned to Daphne Major to collect data on the successive populations of finches. The variation, island by island, of the beaks of these birds was one of the most telling pieces of evidence that Darwin took to support his hypothesis that the mechanism of organic evolution was natural selection. The Grants and their assistants captured, weighed, ringed, and measured the beak of every finch hatched each year as it grew to adulthood. They were thus able to relate the size of the brood in any one year to the characteristics of those birds that survived to breed as adults in the next generation. This was not a statistical survey of a sample of a population. The data were the results of measurements of the entire population of the island, bird by bird. Each beak shape and size of the finches is adapted to a certain size, type, and shape of seed. By counting the seeds on sample areas of the feeding grounds of their birds, they accumulated the data necessary to test the Darwinian hypothesis. Birds with beaks better suited to a certain pattern of edible seeds would tend to have more offspring, thus generation by generation changing the physical shape of the beaks of the whole population when those seeds were abundant.

This was a study of a model of a natural system, the whole biosphere, and the processes that go on in it. It was a study of an analogue of a process that occurs throughout the world of living things. It stands in contrast to the study of evolutionary genetics in captive populations in artificial model worlds but in a different way from that in which Dobzhansky's studies stood in relation to captive *D. melanogaster* genetics.

By a combination of industry, shrewdness, and luck the Grants succeeded. Variation by natural selection does occur and can be seen to occur—indeed remarkably readily.

Two species became the focus of the study. Cactus finches lived in close relationship with the cacti on the island. The fortis finch was larger and roamed more widely through the available plant species. Both finches live mainly on seeds. The seed cases needed to be broken open and the 'meat' extracted. In other places some birds drop nuts from a great height to break them open. Urban crows have learned how to place hard-shelled nuts on the road so that they will be run over by cars and their shells cracked that way. But Galapagos finches crack seed shells with their beaks. The stronger the beak, the tougher the shell that can be cracked. In a year of heavy rains, there are more softer seeds, while under drought conditions both species are soon left with only the toughest seed cases from which to extract a meal. Plainly, one would expect the bigger birds with the larger and stronger beaks to be better fed under drought conditions. Would this affect the relative rate at which these birds reproduced?

The breeding patterns of these species differ in subtle but important ways. Cactus finches pair off and begin their reproductive cycle before the rains come. The fortis finches do so only when the rains have begun. If there is a drought, they do not breed at all. For both species, the number of nestlings in a season and even the age of a female bird for laying the first clutch of eggs also differ with the rainfall. If the rainfall is particularly heavy, the reproductive rate shoots up. Not only do adult females have larger clutches, but also younger females than usual begin to mate and lay clutches of eggs.

Taking all this together, it is clear that flood and drought are likely to be the dominant influences on the distribution of beak types in the population of both species year by year, if Darwinian principles are correct.

I have already mentioned luck. For the first four years of the project little happened. The rainfall was average. The seeds were well distributed among the hard and the soft varieties. The populations of the two species scarcely varied year by year. Then in 1977, while the Finch Team was represented by Peter and Lucille Boag the rains did not arrive on schedule. A severe drought set in that did not break until the end of the year. Many finches died. Only the large males with the deepest and strongest beaks survived in any quantity while the females of the fortis finch perished in great numbers. During the drought the nestling cactus finches all perished, since they had begun to breed before the time at which the rains were due. The fortis finches did not breed at all during the drought.

When the rains did come in January 1978 Peter and Lucille Boag were able to watch at first hand the breeding pattern that powers natural selection. There were six males for every female finch. The role of sexual selection became very clear. The females chose to mate with the larger finches which also possessed the most useful beaks. The next generation of fortis finches was larger than the previous ones with bigger and stronger beaks. This pattern of development continued after the return to normal meteorological conditions.

By ringing and recording the vital statistics of every finch on the island, year by year, a project accomplished by Trevor Price for the 1979–80 season, the relationship between the individual members of the population of each generation to their actual parents could be followed. At the same time, blood samples were taken from every finch in the later years of the project. This treasure trove of material is yet to be fully exploited. The results of the study of the data collected so painstakingly are overwhelming. There is selection of locally advantageous characteristics by natural selection, particularly by sexual selection. Looking at the population individual by individual the Grants and their team were able to show conclusively that the surviving females of the fortis finch picked the largest and best equipped males to pair off with. They also showed that this choice resulted in nestlings that grew into larger and better equipped adults. Darwin's general theory has been vindicated in detail.

A year or two later there was a season of excessive rainfall and a consequential burgeoning of plant life and a plethora of seeds, including lots of soft ones. Once again the population of descendants changed. The population of cactus finches recovered.

The Grants and their assistants watched and recorded a natural experiment. Expressed in terms of the simple independent/dependent variable pattern, the varying weather was the independent variable. The intervening dependent variables were the quantity and type of seed production. The ultimate dependent variables were body size and beak shape of the birds. These were found to vary with the independent variable, the rainfall.

But Daphne Major was much more. It was an analogue of the whole Earth. The two species of finch represent the millions of organic species. The decade of climatic changes represented environmental differences that sometimes spanned millions of years of geological time. The quality of the model depends on the balance between fruitfulness of the similarities between Daphne Major and the rest of the Earth and the scientific irrelevance of the differences between a small island in a great ocean and all the other environments of the Earth.

The material collected by the Grants in their thirty-year study has been used by geneticists at Harvard to identify the genes responsible for the shapes and sizes of the beaks of the finches, and so to track the selection process to its biological roots.

Modelling our world at early times

One problem for which models seem particularly apt is the study of the early conditions on the Earth. Studying suitable models has been much to the fore in the search for the origins of life around 4,000 million years ago. Evidently those conditions are inaccessible. They have long since gone, and so far as we know, the outer layers of the planet, earth, air, and water, have changed greatly in the intervening aeons. However, sometimes an existing natural phenomenon can be singled out and treated as a model for following processes long since extinct over most of the Earth. In cases like this the apparatus which realizes the model in material stuff is not made but selected from what has been found, just as the Grants found and chose Daphne Major.

This was the case when the conditions at deep vents in the deep oceanic ridges were recognized as approximating the conditions that might have existed on the young Earth when earliest forms of life emerged. The discovery of bacteria living without oxygen and flourishing at temperatures that would cook the usual run of living organisms was the first step to singling out these naturally occurring phenomena as models for the conditions of the early Earth.

Though only recently discovered, organisms capable of living under conditions of extremes of heat, cold, desiccation, pressure, and so on have been the subject of massive amounts of research. Collectively they are now known as 'extremeophiles', lovers of extreme conditions of life. Each type of extremeophile has picked up its own unique description—'thermophiles', for instance, thrive in extreme heat.

TOO HOT TO HANDLE

The surface of the Earth thousands of millions of years ago was a good deal hotter than it is now. The processes with which life began are very remote from the biology of the organisms that now fill the biosphere. Whatever form the earliest creatures took, they must have been capable of sustaining life at temperatures that would be fatal to the life forms of today. The discovery of thermophiles at Yellowstone Park by Thomas B. Brock of Wisconsin University in the Yellowstone springs in the 1960s opened up the possibility of a model for the organisms of the young, hot Earth.

Tom Brock has had a distinguished career as a microbiologist, first at Indiana University and for many years at the University of Wisconsin, at Madison. At some time in his long life he turned from the 'indoor' world of microbial cultures to the great outdoors. He became a leader in microbial ecology, the study of the role of bacteria in the natural environment. It was while pursuing these studies that he found the extremeophiles, candidates for a model of the earliest life forms on Earth. On the cooler side he has done a great deal of work on the microbiology of lakes and ponds, the science of *limnology*. Not surprisingly, for an out-of-doors scientist he has spent much of his retirement in the service of the Nature Conservancy, overseeing work on the restoration of prairies and oak savannahs in Southern Wisconsin. His interest in the outdoors may seem surprising for someone brought up in the city of Cleveland. However, his childhood home was near a derelict farm, and there was a 'nature' park nearby where he and his friends went hiking.

From his Yellowstone studies Brock reported the presence of micro-organisms even in boiling water, apparently thriving in temperatures ordinarily used to sterilize medical equipment.[12] These creatures attached themselves to rocky surfaces, extruding long filaments into the boiling pools. They are prokaryotes; that is, they lack a cell nucleus and other differentiated internal structures. And they are not confined to Yellowstone but are found the world over in naturally occurring hot springs.

Other kinds of thermophiles have been found in the ocean, around the vents of undersea volcanic activity on mid-ocean ridges. There both temperature and pressure are higher than on the Earth's surface. These creatures too might be useful as models for the organisms of the earliest life forms.

THE ICE WORMS OF TITAN

The arctic regions of the Earth in the here and now offer another opportunity for model-making by a judicious selection of natural phenomena. Micro-organisms exist on the Antarctic continent under conditions that would deep freeze the usual run of living things. Yet these conditions are not so different from those that space explorations have suggested might exist on the Saturnian moon Titan. Antarctic life forms and the conditions of their existence could be a model for research into the possibilities of life there. Titan, the fifteenth moon of Saturn, is a substantial rocky sphere, with a deep crust of water ice. Its diameter overall is about 5,000 km, not so very different from that of the Earth. It was discovered in 1655 by Christiaan Huygens, his name being an apt choice for the space probe that landed on the Titan surface. Unlike our moon Titan has a substantial atmosphere of nitrogen with a small but significant amount of methane. This accounts for the variety of organic molecules that have been found in the Titan atmosphere. There are clouds, probably of methane. The average temperature is about -155°C so it is likely that it rains liquid methane. There are signs of erosion and lake-like formations. Could there be living beings on Titan?

Following the best scientific procedure the trick is to make or find a model of Titan-like conditions here on Earth where it can be subjected to detailed study and analysis. Enter the ice worms of the Gulf of Mexico. In 1997 an expedition from the oceanography department of Pennsylvania State University found colonies of worms that live in and on mounds of methane ice, created by intense cold and pressure of the deep ocean.[13] This ice usually forms below the surface of the deep sea deposits, but in the Gulf it appears on the ocean floor as large mounds. Thousands of small worms about 2 cm in length live on these mounds, and in the many fissures and natural tunnels that riddle them. The worms were first noticed by Charles Fisher during a submarine dive. Henahan quotes him as remarking that 'these are high order organisms that can live right on methane hydrates.'

Just exactly how these creatures live and what is their relation to the methane ice which they have colonized is still to be worked out. However, here, it seems, is a possible model for the life forms of Titan, if they exist. These creatures seem to use methane as the source of the energy upon which their life processes depend. The nitrogen atmosphere on Titan contains a significant amount of methane, just as the nitrogen atmosphere of the Earth contains enough oxygen to support a rich biosphere of aerobic beings. Here there are two aspects of the 'methane' model. The similarity between the proportions of atmospheric nitrogen to its biologically significant gaseous partner supports one side of the model, while Fisher's worms are models for creatures that use methane as the vast majority of creatures in the Earth's biosphere use oxygen.

Commentary

In this chapter we have examined five 'model worlds' growing in scope in time and space. Clarke's vole worlds were analogous to the grasslands of which these little animals are native. Dobzhansky's *Drosophila* were certainly understood as a model for the genetics of inheritance in whatever species it was occurring. The Grants managed to demonstrate in a readily observable form and in a very confined natural environment the existence of a cluster of short-term examples of natural selection of which the long history of the biosphere is presumed to be made up. The two extremeophile models go yet further back and farther afield. Thermophiles can help us picture life in the hot pools of the early Earth. We can have a glimpse of possible life forms that might exist in the extreme cold that prevails in many distant places, such as the surface of Titan, by taking the worms that live on hillocks of frozen methane in the ocean depths as a model to help imagine organisms native to these conditions.

To understand the significance of this kind of modelling in the growth of scientific knowledge it is important to distinguish the 'logic' of model world building from that of the conventional idea of testing hypotheses.

In some cases, such as those described in the previous chapters, it has proved possible to isolate just one strand of a complex natural process, by contriving to hold all other factors in a situation constant. Scientists sometimes use a terminology derived from the mathematical representation of the relation between causes and effects, the notions of independent and dependent variables, to describe their methods. The experiment can be thought of as manipulating the independent variable, say the pressure, and tracking what happens to the dependent variable, say the volume, when all other factors in the setup such as temperature, are maintained constant. This kind of experiment was taken up by J. S. Mill and many who have followed him as the exemplar of scientific method. By contrast model worlds include many or most aspects of the world of which they are models. The model and its subject are related by similarities and differences; that is, each is an analogue of the other. Since model worlds are better known than the world they model, the growth of scientific knowledge proceeds from what is discovered about the model to claims, all else being equal, about their subjects.

Can a hard and fast line be drawn between running model worlds and conducting Millian experiments? It is perhaps best to see each type of procedure as defining the poles of a continuum of experimental methods. It is possible to carry out Millian experiments with more than one factor being manipulated, and more than one effect being produced. Model worlds necessarily consist of only a selection of the many relevant features of the worlds being modelled.

Chapter 9
Practising Deception: Spring Wheat and Midwife Toads

This book is about the many ways that scientists, honest and dishonest, have made use of equipment at least part of which has been drawn from a living source. Though this chapter is about scientific fraud, a key part of the material needed for dishonest endeavours is just as much drawn from the living world as that put to use in honest research projects.

There are very few examples of deliberate fraud in the history of experimental science. Temptations are subtle and resistance to them not so difficult. However, sometimes the attraction of dishonesty proves too strong. What would lead someone to knowingly fabricate results? Scientists rarely undertake the arduous tasks of experimental work for money. Apart from intellectual satisfaction the reward I suppose is fame and honour. Think of the way we refer to scientific equipment: the *Wilson* cloud chamber, the *Geiger* counter, and so on. Concepts get the same treatment: the *Celsius* temperature scale, the *Hubble* constant. Laws and theories attract the names of their inventors: *Boyle's* Law, *Le Chatelier's* Principle, and so on. One can easily imagine someone saying reflectively 'if only I could get my name on a piece of apparatus, a concept, a discovery or a theory!' Just occasionally a scientist gives way to temptation to try to achieve this kind of immortality by fraud. This is really an extraordinarily risky life strategy, since fame can follow only on publication. But with publication of the results and descriptions of the equipment with which they were obtained comes the risk of the failure of replications undertaken by others and the inevitable unmasking of self-deception at best and fraud at worst.

The material basis of the apparatus or instrument employed for deceptions makes it very difficult to conceal deceit. If the results of an experimental programme come under suspicion it is likely that either the apparatus is itself a fake, or, if it is genuine, the results of using it must have been 'massaged' in some way. Whatever happens, the instruments and apparatus made of glass, metal, and clay or of living material taken from plants and animals must have been manipulated to produce the results which first startled the scientific world. In principle the operations which produced these new and surprising results can be recreated, though sometimes with difficulty.

However, the relevant scientific object involved in fraud can be involved in several different ways. In our first story, the saga of Lysenko and the alleged transformation of the nature of wheat, we encounter real plants and fake experiments. In the second story, Kammerer's long years of efforts to produce a true breeding line of toads by environmental influences alone, we find real experiments and at the end of the day a fake specimen. The experiments did not prove what they were claimed to prove. In one case the experiments had not really taken place. In the other the specimen produced as evidence for a remarkable experimental result was not what it purported to be.

Plants for testing a hypothesis

Trofim Denisovich Lysenko was born on 30 December 1898 in the village of Karlovka in the Ukraine into the peasant family of Denis and Osaka Lysenko. He was the eldest son. As such he was expected to assist his father in the business of the farm. Like Pavlov, though for quite different reasons, his formal education was delayed until he was 13. He learned to read and write at the local school. He was clearly very bright and energetic such that he entered the Poltava Horticultural School in 1913, for a two-year course in scientific farming. The First World War intervened in his life as it did in many of his contemporaries, such as the famous psychologists Lev Vygotsky and Alexander Luria. However, in 1917, the year of the Revolution, he was admitted to the Uman School of Agriculture and Horticulture. At that time the Ukraine was under the occupation of the Germans and it was not until 1921 that he began work in the sugar industry at one of the collection stations. At the Uman School he had come under the influence of the fruit breeder I. V. Michurin, whose anti-Mendelian views were to prove very influential in post-revolutionary Russia. Michurin was famous for his slogan 'We cannot wait for favours from nature; we must wrest them from her.' In 1931 Trofim Lysenko found a post in the Odessa Institute of Genetics. It was here that he began his systematic, but as it turned out, fraudulent experiments on wheat.

The instigator of one of the most notorious frauds in science history was sincere in his belief in the possibility of the inheritance of acquired characteristics, and also, it must be said, to have become a little mad. So passionate was his attachment to the idea that the very nature of living species, particularly plants, could be changed by treating the seeds before planting, and to the development of a crop technology based on this idea, that he massively falsified the results of his various experimental programmes. So potent was his oratory in support of his apparent discoveries that he managed to prevent biologists in the Soviet Union from pursuing research in genetics for nearly twenty years.

The case makes very clear the difference between the morality of politics and the morality of science. Correctness in the political sense is based on the effect of a belief on the status or conditions of life of some group, regardless of its correctness in the scientific sense. Most essays into PC are harmless enough, but in the Lysenko affair, the results contributed to a human catastrophe, the failure of Soviet agriculture and the ensuing famine. The appeal of Lysenko's anti-Mendelian stance to the Party comes out very clearly in a little book produced in 1950 by James Fyfe, a devoted follower of the Lysenko line.[1] Accusing Mendelians of unscientific Aristotelian essentialism he went on to declare that 'the genes are an image of the ruling class as it sees itself.'

Reports from contemporaries show Lysenko to have had that combination of dogmatism and charisma that is essential for a prophet but fatal for a scientist.

According to the Lamarckian Theory, the way of life of a plant or animal in one generation can lead to changes in the genetic material, which can then be passed on to subsequent generations. Changes in the genetic material of a species are directly influenced by the local environment, so the theory suggests. According to Darwinian Theory mutations or changes in the genetic material occur independently of the environment in which a plant or animal population lives. Whatever characteristics the members of a species may have at some historical moment, the role of such characteristics in the processes of reproduction, favourable or unfavourable, determines whether this genetic material becomes more prevalent in the next generation. Mutation, changes in genetic material, and the selection environment are independent of one another.

At the centre of the Lysenko affair was the problem of how to increase the yield of wheat. According to Mendelian genetics, selective breeding would certainly have succeeded in developing high yielding strains more likely to flourish in the 'new lands' across the Urals than existing strains, but the process would have taken years. If Lamarckism were true then treatment of the seed corn this year would have an immediate effect on the yield from the very next crop, or at least one in the very near future. And that change would be permanent. That was what Lysenko set out to prove.

The story begins with his experiments on the phenomenon of *vernalization*. The wheat crop in the Ukraine was harvested from two different strains. Winter wheat was sown in the autumn and, surviving the winter, was ready for early harvest in the following summer. Spring wheat was sown in the spring and matured in the early autumn. In this way the land could be made to bear two harvests. Somehow it was found that if winter wheat were to be allowed to sprout, and instead of being sown then and there, when it would be protected by the soil from the weather, it was subjected to just enough cold to prevent further growth, when finally sown in the spring it matured more rapidly. This is the phenomenon of vernalization.

Young Trofim Lysenko began experiments with the technique of vernalization in the praiseworthy endeavour of increasing the annual yield of the wheat crop. Apparently the technique led to an enhanced yield from the winter wheat sown in the spring. If this were so it could be the basis of a tremendous advance in agronomy. A large-scale test of the process was carried out by Trofim's father, Denis Lysenko. This test was not a scientific experiment—rather a demonstration of the efficacy of a practical farming technique. By subjecting the winter wheat to vernalization, Denis Lysenko did succeed in obtaining a greater yield from a spring sowing of the treated seeds than would have been expected from the usual procedure.

Here began Trofim's slide into dishonesty that would step by step slip into a massive fraud. It seems clear that the younger Lysenko told his father of his work. Denis, on his own initiative, tried out the idea. It seems to have been successful, but then Trofim claimed all the credit for himself! He even claimed to have brought the vernalized seeds from Odessa to Karlovka. Perhaps this bit of self-aggrandisement was not so venal a sin. Did it really matter whose initiative lay behind the demonstration, father or son? So far, the sowing and reaping of just one crop of wheat was hardly an experiment. There were no controls, that is, crops planted with untreated wheat under the same conditions to check whether the vernalization process had been responsible for the increased yield. The results of this trial were lauded in the newspapers and Trofim Lysenko was hailed as a hero. A newspaper article, describing the success of the first vernalization experiments, lauded him as the 'barefoot professor' from a humble background who could succeed in an important scientific project without a university education.

However, there was a second fatal aspect of this event. Trofim Lysenko misinterpreted the nature of the process of vernalization. Instead of seeing it as a technique that accelerated the growth of pre-germinated plants, he believed that the process changed the *nature* of the wheat. This became an *idée fixe*. The slide into fraud followed as he desperately hung on to this erroneous idea, until finally he had eliminated not only the experimental evidence for the genetic account of heredity but its exponents as well.

Spurred on by the erroneous idea that vernalization changed the nature of the wheat, Lysenko began an experiment to produce true breeding wheat strains by seed treatments derived from the process of vernalization. This was a properly planned experiment using controls. For this project he laid out an 'apparatus' very like that used by Gregor Mendel. Lysenko sowed the seed wheat in a number of plots so that seeds could be systematically subjected to different treatments. The results could be expressed numerically as the relative weights of the final crop yield when the wheat was reaped. The seed was sown in four identical plots.

The plots were alike in all relevant respects, so here was a proper experimental test of a hypothesis. The hypothesis was simple: pre-treatment of the plants between germination and growth could change the nature of the seeds. The test was equally simple. If the hypothesis was correct, comparison of crop yields should show that a greater weight of wheat was produced by the treated seeds than from the plots where untreated seeds were sown as a control. Seed sown in the following year, taken from the yield of the treated crop, should give rise to an even more weighty crop.

Unable to wait for the results, so committed was he to this theory, Lysenko announced in advance that he had found a way to dramatically increase the crop yield. His biographer, D. Joravsky, remarks that apart from some early publications all the announcements and descriptions of his 'scientific' work were written by journalists and appeared in the popular press.[2] Unfortunately none of the treatments were successful. Instead of an increased yield the treated seeds actually performed poorly. Having announced the success of the experiment in advance Lysenko fell into a much more serious moral fault than simply claiming to have been the instigator of a field trial. He committed an outright fraud. He made sure that not all the wheat from the untreated plots was weighed. In this way his hypothesis of the efficacy of the treatment could be vindicated. These trials took place at the Odessa Institute, where he had assistants in his experimental work. It would seem from the account that at least some of these assistants connived in the fraud.[3]

But the matter soon turned in a sinister direction. A. A. Sapegin, the Director of the Odessa Institute of Genetics, where the experiment had been conducted, discovered the fraud. According to V. N. Soyfer,

> [He] chanced to notice that some sheaves of wheat had been left unthreshed. At first he thought that some lazy or forgetful assistant on Lysenko's team was to blame, but on closer inspection of all the plots he discovered the truth: The sheaves had been left standing on purpose and only on the control plots. By systematically 'forgetting' to weigh these sheaves in the yield of the control plots, Lysenko's staff had exaggerated the relative yield of the experimental crop.[4]

Soon afterwards Sapegin was arrested and imprisoned as a 'wrecker'. By then Lysenko had become closely involved with the Party, and with the desperate need to find a way of augmenting wheat production quickly.

From Soyfer's account, it seems that at this moment Lysenko hoped that somehow some modification to this pre-treatment of the seeds would be found eventually to vindicate his claims. His belief in the idea that after a few treatments the nature of seeds could be changed was so deeply held that he saw the failed experiments as mere temporary setbacks. He might even have felt that the fraud was a justified manoeuvre to maintain the momentum of the experimental programme. Of course, it is not difficult to put down some of the motivation to his desperate need for fame, for the poor peasant boy to succeed.

Lysenko's malign reign began with a speech to a Congress of Collective Farm Shock Workers in 1935 in front of Stalin himself. Speaking of his own origins Lysenko declared that as a man he was made by the conditions of the Soviet Union. 'And to feel oneself part of such an environment', he said, 'it is something much greater than happiness.' Stalin's response was enthusiastic—'Bravo, Comrade Lysenko, bravo!' On this occasion Lysenko violently denounced Mendelian genetics as 'reactionary and decadent', clashing as it did with the Michurinism to which by now he had become thoroughly committed. Mendelian geneticists were 'enemies of the Soviet people'. According to Martin Gardner, Lysenko claimed that his speech had been approved by the Central Committee of the Communist Party, the secretary of which was Josef Vasliyevich Stalin himself.[5] No doubt that was true.

In 1935, official approval of the man and his projects brought the award of the Order of Lenin, the highest honour a Soviet citizen could achieve. During the next few years Lysenko became more and more immersed in politics. In 1936 he became a member of the Supreme Soviet, the notional governing body of the Soviet Union. Throughout this time he waged a war not only on Mendelian genetics, but on the scientific fraternity that advocated it.

Criticisms of the theory behind Lysenko's experiments came from leading Russian geneticists, particularly his long-time sponsor and friend Nicolai Ivanovich Vavilov (1887–1943). Vavilov supported the extension of the vernalization experiments, but he interpreted them very differently. They did not show a change in the nature of the varieties of wheat but exploited aspects of the conditions under which the wheat plant developed from the first sprouting. Change of the nature of plants is genetic. While continuing to lean on the support of Vavilov, Lysenko began a campaign against genetics and by implication against geneticists. This is the technique of someone who when chided for being late for an appointment attacks the institution of punctuality.

That Lysenko's career of lies and deception should have prospered needs an explanation. From a scientific point of view two grand theories of the nature of biological species were in contention, the Darwinian and the Lamarckian. In the 1920s definitive tests of the theories by direct observation of the genetic material and an understanding of its chemistry did not exist. The apparatus of garden plots, specimen seeds, and comparative weights of crop yields was well designed. The upshot of the conflict between the theories should have been unequivocal. The genotype of a species was not materially affected by the environmental conditions of any single generation. Changes in genotype were spontaneous or caused by some agency other than the immediate environment, such as irradiation. Only the phenotype, the form of the mature plant or animal, was affected by the immediate environment. Epigenetic forces intervened between the hereditary material and its expression in an adult organism. In the 1935 speech that made his name he denounced his Mendelian opponents as wreckers driven by class interests. On the scientific question he declared, 'We recognize mutability of the genotype in the process of the ontogenetic development of plants... [we] can already, by our methods of raising, change the nature of plants in a controlled manner in each generation.'[6] Ultimately, he implied, new species of plants can quickly be produced. As Joravsky points out, this claim could not be queried by the government, since the bureaucrats involved were quite uneducated in all but the official doctrines of Marxist-Leninism.

Stalin's involvement was fatal for the leading biologists in the Soviet Union. Many simply joined the rush to support Lysenko. Those who did not suffered the usual fate of Soviet dissidents. They were dismissed from their posts and sent to labour camps from which many never returned. On 6 August 1940 even Nikolai Ivanovich Vavilov was arrested. He died in one of the camps on 26 January 1943 of neuromuscular dystrophy, probably brought on by starvation. It was not until the 1960s that his reputation was rehabilitated. Even after Stalin's death in 1953 the ukase on Mendelian genetics continued under Kruschev. Had Lysenko by this time become mentally deranged? There is a photograph of him holding up to view the results of a truly amazing transformation—rye into wheat! (See Figure 15.)

Eventually the utter failure of Michurinism and the penetration of Soviet science by ideas from the West had an effect. Lysenko and his projects were officially discredited and finally abandoned in 1965. However, the turning point was the publication in the *Botanical Journal* of an article by Lysenko followed by a vigorous refutation of Michurinism by V. N. Sukachev, the editor. Various claims to have observed climate-driven species change were unravelled and shown to be spurious, such as the claim that a spruce tree had emerged from a pine, and that a hazel tree had grown from a hornbeam. Perhaps surprisingly Lysenko retained his post as Director of the Lenin Hills Agricultural Research Station until his death in 1976 in the relatively palmy days of Leonid Brezhnev.

The *overall logic* of Lysenko's project is well within the boundaries of appropriate scientific work. His vernalization experiments were, like the Grants' close observations of Darwin's finches, *exemplars* for whole classes of phenomena. The garden plots

Figure 15. Lysenko measures wheat.

where the treated and untreated seeds were planted were manageable models representing huge agricultural events, covering millions of hectares. The local logic of the use of the experimental method was also entirely appropriate. The principle goes back to the writings of William of Ockham. It was revived by J. S. Mill in the nineteenth century in the form of the joint method of agreement and difference. If an effect occurs in the presence of some factor, *A*, and does not occur in the absence of *A*, then *A* is the likely cause of the effect. Falsifying the yields simply vitiated the logic. The 'experiments' were worthless.

Looking over at the other side of the laboratory where the apparatus is made of inorganic materials such as glass and metal we also find cases of the suppression of results unfavourable to a hypothesis. Physics and chemistry too have their quota of scientists 'massaging' the experimental results to fit a cherished theory. The fault, so the perpetrator believes passionately, must lie in the inadequacy of the experimental manipulations than in the theory. Theories make things make sense. Casting them out threatens intellectual chaos. In one famous case, the scientist, R. A. Millikan (1868–1953) received the Nobel Prize for his discoveries, whereas Trofim Lysenko earned universal obloquy as a 'crook'. What was the difference?

Gerald Holton made a close study of Millikan's experimental notebooks for the crucial years of 1911 and 1912. Summing up the results of his investigations Holton remarks that Millikan 'quietly selected among his data, using only about 40% of his 140 oil drops [each the basis of an independent experiment] in determining the charge on the electron'.[7] As we have seen Trofim Lysenko also quietly selected among his data using only a proportion of the yields from his experimental plots. While Lysenko failed to get any results supporting his Lamarkian predilections, at least 40% of Millikan's results were in tune with the views of his peers about the constancy and unity of the charge on the electron. It seems he was right not only in his value for the charge on the electron but in his selection of 'reliable' results. Lysenko attacked those who doubted his rejection of Darwinian genetics, persecuted those who criticized him, dreaming perhaps that one day he would achieve the results he had hoped for. Lysenko suppressed the unfavourable crop yields for dishonest purposes. Millikan suppressed the 60% of results that did not fit the hypothesis because he had an unrivalled sense of the way experiments can go wrong.

Millikan's experiment was amazingly simple in principle, but required considerable manipulative ability to manage. Electrically charge oil drops were suspended one by one in an electric field. The force exerted by the electric field on the charged droplet must then be exactly equal to the downward pull of gravity. Using X-rays the charge on the droplet under study was changed, and a change in the electric field was required to maintain it in place against the effect of gravity. Each drop was subjected to this treatment many times. Millikan selected the smallest difference between the initial and the induced charge on a drop as the minimal electric charge, the charge of the electron. Clearly many things could go wrong in such an experiment. Rather than track down the cause of each and every deviation Millikan just scrapped the 'wrong' results.

This is not unusual in physics. The late Kishan Mathur, who discovered the 'apple-shaped' liquid droplet, immortalized by an engraving on the entrance to the Birmingham University Engineering Building, always discarded the results of experiments done early on a Wednesday morning. On Tuesdays he and his girlfriend would stay out till all hours, and he doubted the accuracy of his hand-held camera shots the next day. This was known to his friends as the 'Sweetie Pie' effect.

Kammerer's model of a Lamarckian world

The extraordinary story of the fraud attributed to the Austrian biologist Paul Kammerer involved the attempt to construct a working model of a possible world. In Kammerer's possible world environmental changes in the bodies of certain animals could be fixed as permanent heritable features displayed in a long breeding line. The trait became heritable by repeated exposures to the environmental stimulus generation after generation. In short Kammerer was trying to demonstrate the Lamarckian theory of the inheritance of *acquired* characteristics by another kind of experimental programme.

We have been through all this before with the Lysenko case. However, Kammerer's story is very different. First of all the logical basis of his efforts was quite different from Lysenko's. Furthermore, the detective work done by Arthur Koestler seems to clear Kammerer of any dishonest interference with the key specimen that purported to display the characteristic that had become heritable after a long breeding line.[8] Lysenko was a liar and a cheat. Kammerer was an enthusiast and ultimately the victim of an assistant's misplaced loyalty or perhaps of a jealous colleague's malevolence. The story is all the more intriguing because Kammerer killed himself within a few weeks of the publication of the result of an investigation that showed without any doubt that there had been deception in at least some of the presentations of his studies' results. Was this because of the shame of being exposed as a fraud? Or was it in despair at the ruin of his life's work at the hands of another? Finally Lysenko was not scientifically skilled. His 'experiments' were large scale and conducted largely by the staff of the various institutes he had worked in. Kammerer, on the other hand, was a biologist well known for his expertise in breeding certain amphibians, frogs, toads, newts, and salamanders in captivity. To these matters he brought the finicky perfectionism of one who was also a competent musician and composer.

Let us begin with the experiments. The apparatus for the two series of manipulations that I will describe consisted of water tanks with dry areas which could be reached by the creatures inhabiting Kammerer's model worlds. The models were simplified representations of the natural conditions under which the amphibians that were the main focus of his study lived their ordinary lives. In both cases the animals involved were closely related species, or one should say in hindsight, seemingly true breeding lines that appeared to be distinct species. This caution is necessary since it happens that environmental influences during the development of the members of a species may be such as to lead to two groups of adults being so different in form from one another that for a time biologists classify them as distinct species. The primroses that grow along the banks of the canals in Holland are of the same species as the garden variety but so different in appearance that they were considered separate species. The passage from genotype to phenotype involves all sorts of epigenetic processes in every species. Mostly these are minor, such as differences in the height of human beings caused by differences in diet as well as genetic predispositions. Sometimes they are major, as in the Dutch canal plants.

Amphibians are notoriously difficult to breed in captivity, but Kammerer proved to be very skilled in the necessary procedures and the construction of the right sort of model worlds. The toads that became the centre of the controversy were of the species *Alytes obstetricus*. Unlike other toads this species mates on dry land. The male clasps the female and they remain in this position sometimes for weeks. The skin of toads is dry, so the male has no trouble maintaining his grip. The female eventually lays a long string of eggs which the male then fertilizes. He wraps the fertilized egg strand around his legs where it remains until the young toads hatch. This habit gives rise to the nickname for *A. obstetricus*, the 'midwife toad'. Toads that mate in water follow a similar routine, at least to begin with. The male grasps the female in the same way as does the male *A. obstetricus*. However, in water the toad's skin is slippery and without some additional purchase the male is likely to lose his grip. At the appropriate time the male develops small spines along the edges of the palms of his hands and along the fingers, which at the same time become blackened by the secretion of the pigment melanin. With this extra grip the water breeding toad male stays securely attached to the female until the egg laying and fertilizing is safely accomplished.

The question of the mechanism of inheritance was still open at the time Kammerer began his long-running sequence of experiments. The Darwinian party held that the 'germ plasm', the term coined by August Weismann (1834–1914) for what we now call the chromosomal material in an adult animal or plant, was unaffected by the life events and the environment in which they had lived. Change could only occur through the natural selection of random mutations in the germ plasm. Darwin himself had seriously considered the possibility of Lamarckian processes, and some prominent biologists in the early twentieth century still did so.

With his skill in breeding amphibians perhaps he, Kammerer, might make a contribution to the debate. If there were to be a demonstration of a place for Lamarckian processes in biology, then it would need a stable genetic line, bred through many generations, to show it. If anyone could carry out such long-term studies it would be Kammerer. Members of the *A. obstetricus* species were to be the key components of his experimental apparatus, and his 'Vivaria' model worlds in glass cases were to be *possible* model worlds.

After a breeding programme that had run for many generations in which specimens of dry land breeding *A. obstetricus* were forced to breed in water, and their offspring subjected to the same constraints, Kammerer believed that he had produced males which developed the pigmented spines of the water breeding species of toads. He published his results. Unlike the dubious Lysenko, Kammerer was a well-established scientist working under the aegis of Dr Hans Prizbram, the distinguished head of the Institute for Experimental Biology in the University of Vienna. Animal specimens are dangerously liable to decay unless special care is taken in their preservation. In the end, one particular specimen that had survived under good condition became the focus of the controversy. Here was a male, the descendent of land mating toads, evidently of the species *A. obstetricus*, but with pigmented spines on its hands. Was Lamarck vindicated after all?

The key specimen was examined by a great many biologists, some intent on defending Lamarckism, others on refuting it. This very toad made its way to Cambridge in England where it was subjected to rigorous examination. Kammerer himself gave lectures in Cambridge and in London, where it seems from contemporary reports he created an impression of enthusiasm, competence, and sincerity. The audiences were, for the most part, scientists. However, Kammerer's scientific reputation suffered from his success with lay people—and the sensationalizing of his work by the press. It is fair to say that though many in the scientific community were sceptical of the results, particularly the Cambridge biologist William Bateson, no one doubted either Kammerer's technical skill or his scientific integrity. The midwife toad was a kind of permanent puzzling anomaly. Nevertheless, Bateson and others were unrelenting in their attacks on Kammerer's work. Though unable to prove that the published results were somehow fraudulent their writings insinuated as much.

The suspicion that the apparent heritability of the pigmented mating spines, the nuptial pads, was a deception was established with certainty in 1926. By then the key specimen was beginning to deteriorate. It had been back and forth across Europe and had been scrutinized by a great many people. Perhaps one last close examination would settle its status.

Though Bateson had died the question of the authenticity of Kammerer's results did not go away. In 1926 an American, G. K. Noble, from the American Museum of Natural History went to Vienna for the explicit purpose of examining the exemplary specimen. Along with Dr Prizbram he made a detailed examination of the creature, and, found that the black colouration was the effect of an injection of Indian ink. This was confirmed by the Director's own observations. Of nuptial pads there was now no sign. His observations were immediately published in *Nature* on 7 August that same year.[9] Noble remarks that the absence of the horny spikes on this ancient specimen does not prove that they had never been there. In fact, they had been seen by many people during Kammerer's visit to Cambridge years earlier. However, the faking of the associated pigmentation was indication enough that something was so seriously wrong with the research that any suggestion that it could upset the Darwinian explanation of change by natural selection was empty. The discovery of the fake pigmentation left much unsettled. There was photographic evidence of the presence of the nuptial pads on the original fresh specimen, just as there were sections taken through the forelimbs of some specimens of *A. obstetricus* that showed the internal structure of just such pads.

There can be no doubt that the black pigmentation in the specimen that Noble examined was the result of an injection of Indian ink, a well-known and unmistakable everyday substance. Did Kammerer himself commit this dastardly act? And was his subsequent suicide the result of the guilty shame that overcame him at being found out and publicly denounced as a fraud? Before we turn to discuss the case, following Koestler's analysis, let us examine Paul Kammerer's life and character more closely.

Paul Kammerer was born into the prosperous family of Karl Kammerer, the co-owner of a successful business producing optical instruments, on 17 August 1880. The flux of populations in the Germanic world in the nineteenth century had brought the Kammerers from Saxony to Vienna via Transylvania. Like Sigmund Freud, Paul Kammerer had much older siblings, a half-brother by his father's first wife, and two step-brothers. Paul's mother was Hungarian, and Karl Klammerer her third husband.

Paul seems to have had an idyllic childhood, affectionately treated by his older siblings and cosseted by his mother.[10] Like Konrad Lorenz he developed a passion for animals of all kinds. He seems to have kept some sort of private zoo. A thorough musical education was an essential part of the curriculum for Austrian children. Paul showed a marked talent for music, encouraged by his mother who was an enthusiastic pianist. In many ways Paul Kammerer was typical of upper middle class Austrian youth, well educated and with a broad range of interests and talents. His forays into higher education began at Vienna Academy of Music. However, before completing his musical studies he transferred to the university to study zoology. Nevertheless his musical interests continued, both as a composer of songs and as a member of the inner circle of the musical elite of Vienna at the time. For example he was friendly with Bruno Walter the conductor, and with the great composer Gustav Mahler. One odd story repeated by Koestler has it that the notorious Alma Mahler, after Gustav died, worked for a while as Kammerer's assistant in a study of the praying mantis! Unfortunately Koestler gives no reference for this intriguing story. It fits with other sources which describe young Kammerer as very interested in women, a taste readily indulged in the Vienna of the time.

Paul Kammerer, like his father, was a keen mountaineer. During his trips to the Tatra Mountains he had made collections and observations of amphibians. He had also published on the problems of keeping and breeding captive amphibians. His papers on these matters came to the attention of Hans Prizbram, the Director of the Institute for Biological Research. Having acquired the building and its Vivarium where lizards and other creatures had been on show, from its showman owner, Prizbram needed someone to arrange and adapt the various sections for scientific use. Kammerer took up the post in 1903. He seems to have made an excellent job of the task. There he began the breeding experiments that led him into the saga of the midwife toad. He cleverly created pieces of scientific apparatus out of the glass tanks that had been used to display all sorts of specimens for the public. The total apparatus was completed by the addition of the living component, the toads.

Kammerer took his doctorate in 1904 and in 1905 married the daughter of a noble Austrian family, the Wiederspergs, the members of which, like Kammerer himself, displayed a wide variety of talents. We see him before the First World War in an enviable position in one of the greatest cities of Europe, with an absorbing job and a sparkling social life.

The experiments required a single-minded devotion that perforce had to last for many years. In a lecture quoted by Koestler, Kammerer remarks that 'repeating my experiments is a difficult task; they extended over a span of ten years or more; we would have to wait at least that long for further confirmation.' He collected his specimens himself during walking tours in the country. Throughout this period he published a sequence of detailed reports on the breeding programme and its results. These became very well known and of course, since they tended in a Lamarckian direction, highly controversial. They ran from 1906 to 1919. Eventually he began to publish in the revered scientific journal *Nature*. There it was that the controversy with his Darwinian opponents took fire, particularly as a result of the criticisms by William Bateson.

Throughout all this scientific controversy and Kammerer's sustained efforts to maintain the breeding programme, his private life descended into chaos. His numerous affairs eventuated in divorce from his first wife. His next marriage ended very quickly, and he returned to the shelter of the Wiedersbergs. His bouts of depression and excited activity seem to have been not far short of clinically manic-depressive.

The First World War, over the years 1914 to 1918, was a disaster for the Austro-Hungarian Empire and particularly for its capital city, Vienna. The economic collapse after the end of hostilities was catastrophic. The Institute of Biological Research was hit like every other state-supported undertaking. It was particularly difficult to preserve specimens and it is very likely that the famous toad deteriorated somewhat even though it was apparently well housed. Noble would hardly have been examining a pristine specimen.

This leaves open at least two possibilities as to the perpetrator of the hoax. Some other mischievous acts had been carried out in the Institute by someone described as a 'madman'. Perhaps the scintillating life of the gifted Paul Kammerer had sparked sufficient jealous resentment to suggest a simple plot to destroy him. Subsequent research has shown that if the hoax was to succeed the Indian ink must have been injected very soon before the specimen was to be examined. This suggests another possibility. There is no doubt that the natural pigmentation would have faded over the many years that the specimen had been kept. Perhaps a loyal assistant, without any intent to perpetrate an out-and-out fraud, had thought it good to touch up the specimen to restore its original appearance. The clumsiness of the deception and its incompetent carrying out surely suggest that Kammerer could not have done it himself. This version is made even more likely because within a couple of days he was to leave Vienna for an excellent post in Moscow, to which his equipment had already been shipped. The Russian biological establishment was not as hostile to Lamarckism as western biologists were, as the tale of Lysenko's deceptions showed. Even if Kammerer had felt that his reputation had been damaged in the West there was a hint of Eastern promise. He had written to the Soviet authorities explaining his withdrawal from the post. In that letter he declares that though he did not himself have anything to do with the deception it made it impossible for him to take up the Moscow post.

It has also been claimed that during those fateful days in August 1926 Kammerer had had another piece of bad news. His long string of relationships with women included an affair with Grete Wiesenthal. It seems that Kammerer had been expecting her to come to Russia with him. However, a day or two before he died she seems to have backed out of the plan. Whatever the truth of the matter the humiliation and despair of a life's work coming to less than nothing could have led to the moment on 23 September 1926 when he shot himself on a mountain path. It was only a few weeks after Gladwyn Noble's letter to *Nature* was published revealing the pathetic attempt at deception.

It is perhaps worth a final remark on the whole sad story of the efforts of many people, from scientific bandits to people of the greatest integrity, to prove that acquired characteristics can be inherited. The question could have been settled in our time by any forensic scientist. Just sequence the DNA. No change? Adieu Lamarck!

It would not be right to simplify Kammerer's project into an experimental test of a hypothesis, namely that of the inheritance of acquired characteristics. This would miss the significance of the equipment that he set up. His specimens were taken from the real world of ponds, streams, fields, and mountains. In Vienna he had created a model world at least in outline. But in the real world *A. obstetricus* does not breed in water. By forcing specimens to do so in the Vivarium he created *another* model world different in this respect from the real world from which he had taken his toads. What was the status of this alternative world? The decade-long 'experiment' was designed to show that his Vivarium was a model of a possible world. In that world myriads of species *other than toads* having been subjected to environmental stresses had acquired heritable differences and so become new species. It would have been a world parallel to the Darwinian world for which the Galapagos Islands served as a model. Lamarck's parallel world that Kammerer tried so hard to model successfully is no more than science fiction. It is like one of the strange worlds that Captain Kirk and the crew of the *Enterprise* purportedly run across in their supraluminal dashes across the cosmos.

Turning to the other side of the laboratory again something with a similar logic has happened in the world of glass and metal. 'Cold Fusion', the phenomenon that it seems probably does not exist, is an example of the use of an apparatus constructed wholly of inorganic parts but which has turned out to be, as we say colloquially, 'dodgy'. In 1989 in a laboratory of the University of Utah in Salt Lake City Stanley Pons and Martin Fleischmann were experimenting on the behaviour of palladium when immersed in heavy water. 'Heavy water' is H_2O like ordinary water, but the 'H' is actually 'D', deuterium, an isotope of hydrogen in which the normal single proton of hydrogen is supplemented by a neutron. If an electric current is passed through the system it forces deuterium atoms into the spaces between the palladium atoms. In such close proximity, the Utah researchers believed that 'the nuclei of deuterium atoms would undergo nuclear fusion generating heat and either would be transformed . . . into new elements like helium' or they would disintegrate into subatomic particles.[11] Pons and Fleischmann thought that they had produced more energy than had been put in by the flow of electric current, which could have come only from the fusion of hydrogen nuclei. Instead of submitting a scientific paper they announced the discovery of cold fusion at a news conference. Of course it became a media sensation—here was an amazingly cheap and safe method of solving the problems of energy provision for ever. The military soon became interested because among the by-products that were said to come from the reaction were neutrons and tritium, the third form of hydrogen, the fuel of the hydrogen bomb.

Within days hundreds of follow-up experiments were performed and the phenomena as described just did not reappear. In addition to their claims about the gain in energy during the experiment, Pons and Fleischmann also claimed to have detected the characteristic products of nuclear fusion, neutrons and gamma rays. But this was not supported by evidence either, even it seemed from their own laboratory.

This event looks very similar to the early stages of Lysenko's descent into fraud, and perhaps more strikingly to Kammerer's or someone's futile efforts to save his apparent proof of the inheritance of acquire characteristics in his midwife toads. As Close remarks, 'How did the news media, scientists and government take on board a claim that made no sense within the laws of physics, was based in part on wrong data and that was shown to be flawed within days of the announcement?' The similarity to the Kammerer case becomes more striking when one considers the place of the scientists involved in the scientific establishment. Martin Fleischmann had been born in Karlsbad in Czechoslovakia in 1927, and had been educated in England. He had already had a distinguished career in chemistry in his adopted country and was a Fellow of the Royal Society. Pons was one of his graduate students. Kammerer was a well-respected biologist when he began on the fatal project of the midwife toad. The case on cold fusion is not yet closed—further research is going on in the hope of identifying the phenomenon or something like it.[12] One thing is certain though—Pons and Fleischmann may have been unwise in their premature announcements, but they were certainly not dishonest.

Commentary

The rarity of what one might paradoxically call 'genuine scientific fraud' is a consequence of the scientific method itself. The constant interplay between theory and experiment, and the often neglected but crucial role of the equipment in experimentation, means that science is public in a very demanding sense. Metaphorically, the invitation is always open to come into the lab and play with the equipment oneself. Self-deception is another matter. No one has cast any doubt on the honesty of the people who thought they had discovered cold fusion, nor on Robert Millikan's integrity. However, there can be no doubt that Trofim Lysenko eventually dishonestly manipulated results in the face of the failure of his experiments. At the beginning it seems that he believed so enthusiastically in the Lamarckian possibilities of vernalization that adjusting results to keep the project going might have appeared a good, even noble thing to do. As the years passed and the social network of the Party bound him more and more tightly surely he must have known somewhere in his mind that his declarations that rye had been 'vernalized' from wheat was just not true, and yet his situation forced him into these declarations. Had we joined Alma Mahler in Kammerer's laboratory we would have witnessed a systematic and technically skilled decade-long breeding programme. What if he had managed to produce toads with breeding spines and melanized patches? The evidence of the famous specimen is equivocal and the meaning of his suicide unclear.

Could Kammerer's breeding programme have produced the changes in the 'hands' of his toads? Inheritance has turned out to be a much more complex process than merely passing on a genome. All kinds of epigenetic processes intervene. None of these suggest the Lamarckian theory in any simple form—but the pattern of relationships between environment, genome, and development processes is complex. If there are 'Lamarckian effects' they must be subtle and last for only a few generations.

Could cold fusion actually occur? Is an interaction between close-packed deuterium nuclei physically impossible? Most of my colleagues are unwilling to declare outright disbelief in its possibility, however unsatisfactory the Pons and Fleischmann claims proved to be.

Chapter 10
Inventing Novel Beings: An Imaginary Cat and Virtual Life Forms

We do not always have to use real pieces of equipment to do experiments. Sometimes it is enough to imagine a situation and think out what might be the outcome of manipulating the things with which we have populated it. Some of the greatest scientists have been the most adept at the art of 'thought experiments'. To test out the viability of his conception of absolute motion in absolute space, Newton imagined a pair of globes connected by a string, rotating around their common centre of gravity in 'an immense vacuum'. He concluded that even if there were nothing else in the Universe, an observer could know whether the globes were really rotating because there would be a tension in the string if they were. The only frame they could be rotating with respect to in this situation would be absolute space.

The history of physics is rich in thought experiments. Sometimes a well-conceived experiment plays a key role in moving the science forward by establishing the reasonableness or even the inescapability of a new point of view. For example, Galileo asks us to imagine a horseman at full gallop throwing up a spear above him. Can he catch it again? By reflecting on this image we can see that relative to the horse the spear just goes up and down, but relative to the ground it keeps pace with the horse. It already had the forward motion of the horse in the hand of the horseman. Clearly, the rider can catch the spear. Galileo's analyses of the motion of projectiles and other bodies moving in various ways under gravity were based on the intuition that the vertical and horizontal aspects of a thing's motion under the influence of gravity can be considered independently of one another.

Erwin Schrödinger's thought experiment has become a fixed point in discussions of the meaning of quantum mechanics. What is the status of the *possibilities* for obtaining definite experimental results before an experiment is performed? The thrust of the question depends on the fact that so far as we know the limits of the predictability of events in the world of the very small goes no further, and some would say, could go no further than a range of possible results for any given experimental setup. Each possible result can be predicted with a certain probability, but none can be predicted with certainty. In the world of middle- to large-scale things the results of experimental manipulations can be predicted with certainty. 'Microphysics' or 'quantum physics' deals in probabilities while 'macrophysics' or classical physics deals in certainties. After four hundred years of refinement the concepts of classical physics have became very well defined and the laws in which they figure have been well established. Space and time, mass, energy, momentum, and the concepts of electromagnetism have provided a solid basis for describing and explaining events in the material world.

The past hundred years has seen the opening up of the question of the nature of the minute entities that classical physics presumed were the basic constituents of the material world.

Atoms were revealed as structures of subatomic entities, to which the laws of classical physics cannot be applied with the same degree of precision as they can to large-scale material objects. Most physicists have tried to imagine *models* of micro-reality in which the concepts of classical physics still play a fundamental role, though new *laws* of nature seem to be called for in the theories in which they are applied. On this view the micro-world could be thought of as if it consisted of things moving through space and time along well-defined trajectories interacting with one another according to laws similar to but not identical with those known from the study of large-scale objects. Electrons were endowed with mass and velocity as well as electromagnetic properties such as electric charge. When unexpected effects needed to be explained physicists helped themselves to more of the repertoire of traditional mechanical properties to enrich their imaginary models. Electrons were imagined to spin on their axes, clockwise or anticlockwise.

Erwin Schrödinger was opposed to this haphazard model-making. He wanted to show the absurdity of projecting the concepts of macroscopic physics onto the unobservable micro-world even if only in the form of working models. Unlike many twentieth-century physicists unwilling to give a realist interpretation to microphysics and contenting themselves with refining a picture or model to facilitate mathematical treatments of deep phenomena Schrödinger believed that by trying to forge new concepts it should be possible to give an unparadoxical account of the micro-world that he held must lie behind the world as we can experience it. We cannot observe it, so we must study it with the help of thought experiments. He wanted to move from models to the world that they were constructed to represent.

Biologists have also been ready to perform thought experiments in the absence of the conditions under which a real experiment could be performed. Darwin's account of organic evolution and the origin of species is for the most part a thought experiment. As we saw in Chapter 7 the realization of an experimentally or near experimentally induced evolution of organic forms has been achieved only in the middle twentieth century in the studies of captive colonies of drosophila. By shrewd selection of a restricted and surveyable natural environment the Grants with their 'island as world model' were able to study naturally induced selection processes. Darwin had no such opportunity. The creatures of the Galapagos Islands were open to him only in the very sharply demarcated moment of the present during which he made his observations.

What if one could create an artificial world in which imaginary animals could be seen to evolve? This is the project that Richard Dawkins described in his book, *The Blind Watchmaker*. Dawkins's 'creatures' appear in animal- and insect-like forms on the computer screen. Constraining their reproduction generation by generation for certain desirable properties results in these 'biomorphs' evolving into eye-catching, apparently familiar animate forms. They are not real organisms at all. Yet they can be used to demonstrate the key Darwinian idea of selection of the individuals most likely to survive in a given environment by extending the principle from the deliberate choices of stock breeders to natural selection. The next generation of Dawkins's biomorphs are selected by a repeated application of a particular rule. In this way initially primitive forms can evolve step by step into shapes that may seem to us almost to be designed, though the environment they fit is abstract and mathematical.

How many lives has an imaginary cat?

Erwin Schrödinger popularized one of the most striking, yet one of the most enigmatic thought experiments in the history of physics, the cat in the sealed box.

He was born on 12 August 1887 in Vienna, then the capital of the Austro-Hungarian Empire. Not only was Vienna the administrative centre of the Empire but it was also the scene of a cultural ferment out of which came many of the people and the ideas that shaped European civilization for at least a century.

The connection with Britain and the English language came very early in Schrödinger's life. His mother insisted on his speaking English with her, even on their country walks. As he remarks, this opened up the world to him, so that he never felt a stranger anywhere.

As an adult, Schrödinger seems to have been engaged in a restless and ruthless pursuit of various personal goals and satisfactions. The story of his academic career tells us of a man willing to drop out of a post in which he had scarcely arrived when the chance of a better one came up. Though married for many years to Annemarie Bertel he made no secret of his affairs. In his *Autobiographical Sketches* of 1960 he remarks:[1]

> I must refrain from drawing a complete picture of my life, as I am not good at telling stories; besides, I would have to leave out a substantial part of this portrait, i.e. that dealing with my relationships with women.

Where did the intensity, persistence, and originality of his struggles with the problems thrown up by the discovery of the quantum phenomena come from? He was not so much driven on by ambition as drawn towards ever receding culminations and consummations in science and in his personal life. After serving in the Austro-Hungarian army in the First World War he shuttled between the universities of Vienna, Jena, Stuttgart, and Breslau. His first permanent job was in Zurich. From thence he moved to Berlin. The same year in which he received the Nobel Prize, 1933, he returned to Austria. In 1939 he left for England. His moves in the 1930s were occasioned by his intense dislike of the Nazis, though as an 'Aryan' they posed no personal threat to him.

Later in life, partly as a result of his friendship with the Irish President, Ramon De Valera, himself a physicist, Schrödinger settled in Dublin. The Irish Government created a research institute more or less just for him. Visitors came from all over the world to work there. However, by 1956 he was back in Vienna, teaching physics. He died there on 4 January 1961.

The sixteenth and seventeenth centuries saw the establishment of an ontology or frame of fundamental concepts defining the then scientific view of the world. Though never wholly undisputed it was widely taken for granted for the next three hundred years. So deeply entrenched did it become that it seemed to define the very being of nature itself. According to Robert Boyle (1627–91) and Isaac Newton (1642–1727) the material universe consists of myriads of minute particles, clumping into larger material things and in various states of motion in a vast empty space, the void. The basic entities of the sciences of physics and chemistry were taken to be minute versions of the material things of the ordinary world of everyday life, though at the deepest level they were indivisible. The ultimate corpuscles could neither be supplemented by fresh creations after God's original act, nor could they be destroyed by decomposition.

During the nineteenth century a consensus developed among the members of the scientific community that the material world consisted of two very different kinds of 'stuff', matter and radiation. Matter was atomic, and its behaviour, from very small to very large could be described and predicted using the laws of Newtonian mechanics. The basic properties of material beings were mass and velocity of translation and rotation. Radiation was wavelike, and propagated continuously. Its behaviour could be described and predicted using the laws worked out by James Clerk Maxwell. The basic properties of waves were wavelength, amplitude, and velocity of propagation.

It seemed to many people at that time, not only scientists, that the laws of mechanics and the laws of electromagnetism taken together and applied each in its own proper domain provided a complete basis for all possible future science. Einstein's Special Theory of Relativity upset this neat consensus by forging a unified theory of the propagation of matter and radiation. However, at about the same time, the accepted ideas about the nature of radiation, light, heat, radio waves, and so on were being undermined in another way. The relation between the spectrum of radiation emitted by a *black body*, that is, a source that emits radiation of all wave lengths, and the temperature of the emitting surface could not then be represented in a single law over the whole temperature range. The name 'black body' comes from the fact that something which absorbs all wavelengths of electromagnetic radiation will appear black; that is no light is reflected. Such a surface is also a perfect radiator. Physicists tried to find a single law that would relate the wavelength of the radiation emitted to the temperature of the body emitting it. Wien's Law, for example, was adequate for short wavelengths but failed to describe the relation for long wavelengths. Max Planck (1858–1943) proposed that the radiation should be treated as if it were generated by oscillators that were quantized, that is, 'vibrated' at a range of discontinuous frequencies. Using this assumption it turned out to be possible to formulate just one law relating the distributions of wavelengths in the emission from a black body to the temperature of the body which emitted it.

In another part of the forest, as it were, another discovery was made that upset the seeming divide between atomic and wave phenomena. J. J. Thomson had identified elementary material particles of a certain mass and velocity that carried negative electric charge. It seemed that the corpuscularianism point of view had received further vindication. These minute beings seemed to be candidates for the ultimate constituents of matter. But to which of the great divisions of nature do they belong? Are electrons like bullets? Should we say that *an* electron was emitted from a heated wire and *reached* an atom in a silvered screen after so many nanoseconds? Making sense of this way of talking presupposes that there is such a material being as 'an electron', which is self-identical throughout the motion from source to screen. Even to use such a phrase as 'throughout the motion' incorporates the corpuscularian view point.

Though electrons sometimes behaved like projectiles, there were plenty of experimental arrangements in which phenomena seemingly suggesting a particle nature actually hinted at an underlying reality that was wave-like. When a large number of electrons are passed through a narrow aperture the patterns of light and dark bands that appear on a screen beyond suggest the interference of waves, crests, and troughs amplifying or cancelling one another. Many other similar phenomena were discovered. How could the basis of physical reality be both particle-like and wave-like?

The neat segregation of the universe into discontinuous atomic entities and continuous wave fronts had clearly broken down.

Could a new mechanics be created that would cover both traditional domains of phenomena, and at the same time account for the new experimental discoveries? If everything was quantized, came in blips and bits, how could there be continuous waves?

The positivistic response to this hint of a deep incoherence in the nature of the world was simply to abandon hypotheses about 'ultimate realities' in favour of statistical descriptions of patterns among observable phenomena. Werner Heisenberg took this way out. According to the prescriptions of the philosophical doctrine of positivism, the only reliable and certain basis for any understanding of the world must be confined to the results of observation and experiment. Theories are not insights into the depths of nature but devices for making predictions. Heisenberg's mathematical quantum theory left the underlying processes in limbo, treating the task of the physicist as going no further than creating a formal system that would account for all the known experimental results and enable successful predictions to be made.

Schrödinger took a very different line. As we have seen, classical physics presented two different but complementary pictures of physical reality, the atoms of the mechanical world picture and the waves of radiation. The trouble seemed to have arisen from the need to forge a common theory from two pictures which had been exclusive but complementary. Schrödinger's response was to eschew the positivistic retreat to a mathematics of experimental laws by rejecting a particle scheme altogether. Only waves were real. The wave phenomena revealed by experiments reflected the true material reality. The particle phenomena were artefacts of experimental arrangements. His mathematical quantum mechanics was a wave theory. Not surprisingly, it came to be called 'wave mechanics'. Whether its famous cluster of equations were really about waves became a matter of controversy.

All it needed now to complete the physical picture of the world was a technique for discovering the differences between the fundamental particles that accounted for their arriving at one point rather than another. Gambling depends on there being an indeterminacy in the environment of the casino in the punter's knowledge of the outcome of such things as spinning a coin, the track of a rolling ball, the fall of the cards, and so on. Without impossibly detailed knowledge of the conditions of tossing this coin at this moment, heads and tails have equal probability of coming up, namely 0.5. This is reflected in the fact that roughly equal numbers of each outcome appear as the number of coin tossings increases. However, physicists are confident that if they could measure every aspect of any actual coin as it spins and take into account the influence of the environment in their calculations the result could be predicted with certainty. These extra bits of information that the average punter does not have access to are the values of 'hidden variables'.

After decades of effort by theoreticians and experimenters it has been conceded that there are no discernible differences in the properties of the electrons that fall in this or that region of the screen. The same conclusion has been generalized across the board in subatomic physics. There is, it seems, a basic randomness in the make-up of the Universe. Hidden variables cannot be added ad hoc to quantum mechanics without wrecking the rest of the system.

Albert Einstein found this anathema. 'God does not play dice', he famously remarked. Quantum mechanics, he declared, is incomplete. He expressed dissatisfaction in the technical language of physics in a famous thought experiment with which he confronted the physicists of the times, notably Niels Bohr and Erwin Schrödinger. The Bohr/Einstein debate is very well known, but since Einstein evidently did not understand what Bohr was trying to say, their discussion is frustrating to read. A shootout in which two men of genius fire past each other leads nowhere.[2] Less famous but more constructive and more fruitful in working out the best interpretation of the mathematical demands on the physical theory of the constituents of matter and radiation is the correspondence between Einstein and Schrödinger. The 'cat paradox' comes out of this exchange of letters which served as the basis for a well-know sequence of papers that Schrödinger eventually published as a continuous narrative.[3]

Schrödinger was determined to interpret his 'waves' as real undulatory processes. However, before his program of theoretical development could be carried through, Max Born came up with a way of saving the particle interpretation as a general metaphysical foundation for physics. Schrödinger's wave equation could be treated as a way of working out the probabilities of particles having certain physical properties, for example being at a particular location. Einstein, disappointed in his inability to arrive at a physics of continuous fields, seized on Born's interpretation. Schrödinger never accepted Born's proposal, though it has become scientific orthodoxy. Schrödinger declared that he was opposing not a few special statements of quantum mechanics/quantum theory as it is held today, but, as it were, the whole of it, including Max Born's probability interpretation.

Once Schrödinger's comprehensive wave *interpretation* had been abandoned, quantum mechanics was developed as a comprehensive theory of subatomic *particles*. Schrödinger's wave equation was reinterpreted so that its solutions for any particular values of space and time variables appeared as a string of terms, each of which represented the probability that an electron would be found at a certain place. A new corpuscularianism had almost arrived.

In the original *picture* or model an electron is a thing, in motion along a certain trajectory, of a certain mass and electric charge and orbiting a nucleus for the most part and spinning on its axis. Some electrons spin in one sense and some in the other. This is spoken of as 'spin up' or 'spin down'. It is represented by the 'fourth quantum number'. One can picture this metaphor as the opposite directions in which the thumb of one of one's hands points, with respect to the direction of spin represented by the clenched fingers. Turning the fist upside down not only changes the direction of spin of the fingers but now the thumb points downward. All this talk of spin must be understood to be in the context of a *picture* of a world of minute things spinning on their axes and orbiting the nuclei of their atoms, just as planets spin on their axes and orbit the Sun. However, the spin of electrons is actually an electromagnetic property. It refers to a magnetic moment that would arise as a consequence of the spinning of electrons as minute electrically charged bodies. This field is proposed to account for a strange feature of the electromagnetic spectrum of incandescent gases—the main lines of the spectrum when resolved very sharply are doublets, two lines close together where the early spectroscopists saw only one. This effect is due to the magnetic field of spinning electrons interacting with the field created by their imagined orbital rotation around the nucleus, the source of the main lines in the spectrum. This field is also quantized so that in the particle picture, electrons can occupy only certain orbits. There must be two of these spin fields oriented in opposite directions to account for the appearance of two lines. Hence, thinking back to the original picture of spinning planets there is spin-up and spin-down. The probability that any electron will have one or the other of these field orientations is 0.5.

However, this may look like a new corpuscularianism, but in one respect it is very different from the old. The new theory incorporates the idea that individual electrons could not be marked in such a way as to make sure that one was dealing with the same electron again. To deny that corpuscles have individual identity, as does the new physics, is to go very deeply into a new and radical point of view. As Schrödinger put it, 'This means *much more* than that the particles or corpuscles are all *alike*. It means that you must not even *imagine* any one of them to be marked—"by a red spot", so that you would recognize it later as *the same*.'

This is very easily explained if each 'particle manifestation' is a new product of the running of an experiment a second time. It may be easily explained but it is not easily believed. So much of the entrenched past of physics must be given up. 'In principle, a measurement is a "Procrustation" [A reference to the bed of Procrustes, who "adjusted" the length of his guests to fit the bed]; it produces the value that it [an]nounces as its result, it does not really find it in nature.'

Schrödinger uses the story of the 'cat in the box' to show that it is a mistake to think of particle attributes as microphysically real. The cat story is meant to persuade the scientific community that the particle picture cannot play any role as a model truly representing microphysical reality, rather than just an image to facilitate calculations. Particle attributes like position and momentum are sharp and determinate, but when we try to use these concepts to describe the micro-world they cannot all be determinate at once. In a famous metaphor, they are said to be 'smeared'. There were two tasks to be carried through in this situation. The first was technical—to construct a mathematical theory which would consist of a small number of axioms from which descriptions and predictions for both atomic and wave phenomena could be obtained. Mathematicians such as Werner Heisenberg, Paul Dirac, and Erwin Schrödinger invented several successful systems. The second task was physical—to provide an interpretation of the mathematical systems which would present a unified and consistent picture of the nature of the processes responsible for the experimental results.

To understand the 'Cat Paradox' we need one more fundamental idea, that of a superposition of states. According to the popular interpretation of quantum mechanics, the wave equation describing the evolution of a system can be solved for any particular situation as a series of terms, each of which assigns a probability to a certain value of the relevant mechanical property of a micro-particle being obtained in an experiment, such as position or momentum. Let us take spin as an example to illustrate the paradoxical aspects of using macro-concepts, the concepts of classical physics, in describing the micro-world that is presumed in the electronic theory of the atom, and that appears in the related explanation of the double lines in the spectrum of light given off by an incandescent substance. In an experiment an electron can display one of two spin states, 'up' or 'down' (Figure 16).

Before the experiment the electron to be experimentally examined is said to be in a superposition of states, spin up and spin down. The very idea of a superposition only makes sense if there is presumed to be an electron or electron precursor to which the superposed states can be ascribed. But how can this being have both states at once?

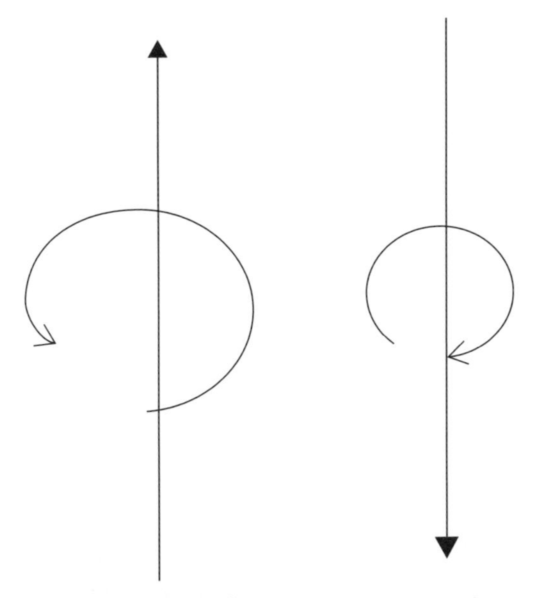

Figure 16. The spin-up and spin-down image.

The idea of superposition can be examined from two directions. Does it make any sense to amplify up the superposition idea from the micro- to the macro-scale? Or does it make sense to shrink down the scale and use macro-scale concepts in a situation that corresponds to the micro-setups of quantum mechanics? Since physics is built out of macro-scale concepts, admittedly according to rather different rules for micro-scale applications, we can ask whether it is proper to use macro-concepts to construct a plausible picture of the world at the micro-level. Remember that experiments yield results that make use of classical concepts. Electrons display highly particulate presentations when experimentally detected. A screen for detecting electrons 'lights up at one instance and at one spot' as Schrödinger remarks.

> Here is Schrödinger's Cat Paradox.[4]
>
> A cat is penned up in a steel chamber, along with the following diabolical device (which must be secured against direct interference by the cat): in a Geiger counter there is a tiny bit of radioactive substance, so small, that *perhaps* in one hour one of the atoms decays, but also with equal probability, perhaps none; if it happens, the counter tube discharges and through a relay releases a hammer which shatters a small flask of hydrocyanic acid. If one has left the entire system to itself for one hour, one could say that the cat still lives if meanwhile no atom has decayed. The first atomic decay would have poisoned it. The ψ-function [that is, the wave description] of the entire system would express this by having in it the living and the dead cat (pardon the expression) mixed or smeared out in equal parts.
>
> It is typical of these cases that an indeterminacy originally restricted to the atomic domain becomes transformed into a macroscopic indeterminacy, which is then *resolved* by direct observation. [After all, an observed cat is either alive or dead!] That prevents us from so naively accepting as valid a 'blurred model' for representing reality... There is a difference between a shaky or out-of-focus photograph and a snapshot of clouds and fog banks.[5]

What is the situation before the box is opened? If this situation were to be a macro-model for the micro-world conception of superposition of states, say spin up and spin down, we would have to say that there was a superposition of states, cat dead and cat alive. Opening the box brings one or the other into being. This is absurd. Releasing cyanide will kill a cat in a split second. Before we open the box there is a probability 0.5 *from our point of view* that we will find a dead cat or a live one because we cannot predict whether the radioactive trigger will discharge. Yet, inside the box, the cat is definitely in one state or the other. Superpositions are conjunctions of possibilities, not of actualities!

In commenting on this situation Schrödinger remarks that 'reality resists imitation through a model', meaning a model that makes use of the existing concepts of physics. If we try to read back from a determinate experimental result to a determinate but unobservable prior state we find that all we have are possibilities—it could have been this and it could have been that! We cannot open the box without doing the experiment! Commenting on the temptation to slide into positivism, Schrödinger says, 'So one lets go of realism and leans directly on the indubitable proposition that *actually* (for the physicist) after all is said and done there is only observation, measurement.'[6]

The upshot of Schrödinger's persistent wrestling with the problem of creating a coherent physics of the micro-world is this: There are irresolvable anomalies for any generalized particle picture of deep reality if we try to create it out of the concepts of classical physics however refined and however segregated into subgroups by the principles of quantum mechanics. The macro-world displays nature predominantly in the form of clusters and structures of separable parts. If the working models of how the micro-world might work draw on classical concepts from the description of the macro-world, absurdity follows. A superposition of the states 'dead' and 'alive' is nonsensical. In just the same way a superposition of the states 'spin up' and 'spin down' as states of an electron, before the spectrum of an incandescent sample is closely studied, is equally absurd. These are not models of the ultimate material processes that underlie these displays. Whatever electrons are, they are not like individual felines, large things! Nor are they little things either. We should not think of them as things at all. This is the point of the Cat Paradox.[7]

Here we have an organic being as the focus of a famous thought experiment. Happily, even if the radioactive sample decays and the cyanide is released an imaginary cat is only 'imaginarily' dead.

Dawkins and the biomorphs

As Richard Dawkins presents his thought experiment in *The Blind Watchmaker*, the imaginary creatures are at best animal-*like*.[8] Furthermore they are part of an exposition of the Darwinian theory of organic evolution rather than the basis of a research level thought experiment, such as Darwin's way of imagining nature as something like a vast farm and garden. The point of the whole thing is to make evolution of new species by the selection of breeding pairs which display minute variations in form and function entirely plausible. However, the biomorph project is set in a more controversial theoretical background than Darwinism in general. It is also an illustration of Dawkins's particular contribution to biology, the clarification of the distinction between Replicators (genes) and Interactors (adult organisms shaped by those genes via epigenetic processes of various kinds).

Dawkins, a zoologist and Fellow of New College, Oxford, is noted for widely read and influential books, proselytizing for the Darwinian point of view in much the same vigorous and entertaining way as had Huxley in the nineteenth century. He shot to fame with his first book, a superb presentation of the Darwinian theory of the origin of species by natural selection, *The Selfish Gene*.[9] This book brought out the full power of the 'Modern Synthesis', the grand theory that brought Darwinian biology, Mendelian genetics, and the biochemistry of DNA, the genetic material, into a fully coordinated synthesis. Furthermore, though the style was popular the text was also a contribution to the general theory of the origin of species by natural processes. Dawkins has been unsparing in his criticisms of Creationists and other opponents of the Darwinian view of nature—just as Huxley was Darwin's bulldog, Dawkins has been declared to be Darwin's Rottweiler.

Scientific research is a many-faceted project. Not only does it rest on the results of experimental practices and well-informed observations, but it also requires a constant attention to the conceptual systems, the working ideas, with which experiments and observation are planned and their results interpreted. To paraphrase the great philosopher Immanuel Kant, 'Experiments without concepts are blind and concepts without experiments are empty.' Science develops as much through the discussion of and even controversy over the appropriate range of concepts appropriate to some field of study as it does through laboratory and field work. Ways of convincing fellow scientists and all those who have an interest in the sciences that a certain conceptual innovation is worthwhile is at least as important as experimental testing of hypotheses or the running of models of natural systems in the laboratory. At the same time, a clear and elegant model of a core natural process is also a tool for spreading knowledge of science throughout the community.

This is the project that Dawkins undertook in creating a world of imaginary beings, the world of the 'biomorphs'. These computer-generated beings could be made to evolve into more and more complex forms by selection, generation by generation, of minute changes in their generating algorithms, according to some arbitrary rule, in a way that represented the methods of artificial selection practised by animal and plant breeders. We will see how Dawkins set about using the biomorphs to illustrate just how powerful the Darwinian evolutionary process is, in whatever medium it operates.

The Darwinian theory of evolution by natural selection is a grand overall conception of the origin of species but its details were not settled by Darwin himself. Dawkins makes very clear that Darwinian evolution is not a random process. New forms of organic beings are produced by *cumulative selection*. Compare this with 'sieving', a one-step process which can, for example, select small stones while eliminating large ones. Repeatedly sieving the same sample leads nowhere. Cumulative selection operates generation after generation with the products of one generation fed into the process anew, so that imperceptibly the members of the ever renewed population shift towards some condition far from the state of the creatures that originated the sequence of parents and offspring. For example, it is advantageous to be sensitive to light. Mutations that favour such sensitivity can accumulate over millions of generations so that a sequence that begins with organisms that react only to light and darkness ends with creatures with the sophisticated eyes of primates. Somewhere along the line of mutations a general sensitivity to light became localized in a light-sensitive patch. Now directional sensitivity became possible. And so, the accumulation of advantageous mutations in the final product—eyes with corneas, retinas, lens, and so on—can give all the appearance of the work of deliberate design.

Let us look more deeply into this process. What are the units of evolution? At first sight it seems obvious that they must be adult organisms, since it is these that survive or perish in interaction with the environment. For Dawkins they are genes. What are 'genes' that they should be units of evolution? A strand of DNA consists, as we all know, of a double helix of molecular units, the order and composition of segments which are the key to heredity. Genes are physically sequences of such molecular units.

What happens when genes replicate? We need two fundamental conceptual distinctions to understand the process. The first is the distinction between the form or structure of something and the material of which that thing is made. This is an ancient distinction and played a major part in Aristotle's paradigm for scientific research. It is still fundamental to scientific research. The second is the distinction between type and token. The same form, in the sense of 'type', can be made concrete as a token many times over each time in different matter. The US quarter had exactly the same shape and symbolic inscriptions for a very long time though millions and millions were minted. All those quarters were the same, but each particular was but a token of that type.

When genes 'replicate', the process is like creating additional tokens of the same type. That type is not something concrete and material. It is the molecular structure of the strand of DNA. Nothing material is passed on. One is simply copying one token of the type as another token with the same structure. How is that possible? The same structure is being realized in different matter.

Glancing at the natural world we seem to see animals and plants everywhere replicating themselves as more animals and plants of the same or slightly different forms. However, this is a very superficial view. The next generation of adult organisms is created by development from a pair of gametes, which have come into being as tokens of the geno *type*. Each is a replica of the type of DNA from which it is copied. Organisms reproduce but do not replicate. An offspring resembles its parents only in so far as it is the end product of a developmental process in which the hereditary genome, formed from unique combinations of the genes of both parents, is fulfilled in some new matter. Evolution is about changes in the 'gene pools' of the population.

Dawkins's argument is presented through a string of powerful and evocative metaphors, at the head of which is the metaphor of the title of his book, the 'selfish' gene. The idea is very simple but was soon almost gratuitously misunderstood. The *process* of organic evolution is nothing but the relative rates of replication of genes, heritable structural units realized in different material, different tokens of the same or nearly the same type. Only secondarily did the reproduction of the adult *organisms* resemble the way tokens are drawn from types. It is rather like the situation in which one admires a house, asks for a copy of the plans and has another identical one built. The blueprints are like genes of which the successive photocopies are their material carriers, each of which might eventuate in a house.

Looking at life over time one can identify two streams of beings. There are the adult organisms, populations of which succeed each other as the years roll by. These beings interact with the environment and survive or perish, reproduce or remain barren, depending on their anatomical and physiological attributes in relation to the actual environment. These are the interactors.

However, there is another stream of beings. Interactors reproduce only through the pairing of minute gametes derived from the sexual cells of the adults, which fuse in the reproductive process to give rise to new beings. These are the Replicators. Only Replicators give rise to Replicators. Whether they do so depends on the success or failure of the life course of the Interactors.

Here was a very clear and even dramatic way of presenting the Darwinian theory of natural selection.

Continuing the story in a very simplified way the selection process could be seen in genetic terms too. Thus a certain gene, when expressed in some anatomical, physiological, or behavioural aspect of an adult organism, was more likely to be inherited by members of the next generation if the biological characters that expressed it led to greater reproductive success of the organisms that carried it. This amounted to an increase in the number of organisms that carried the gene, and so to an increase of the number of instances of that gene in the population.

The metaphor of the 'selfish gene' was Dawkins's attempt to make clear that the upshot of any particular selection process was the spread of the gene that lay behind the reproductively advantageous behaviour of adult organisms shaped by it in conjunction with the rest of the genome. It was 'as if' a gene looked only to its own multiplication. Another metaphor is involved in this story—the gene as an entity that persists from one generation to the next. Of course the actual molecules that form the gene as a thing do not persist through the populations of successive generations—only their structure, or in other words the 'information' that they carry. One writes an e-mail message which exists as a string of on/off switches in a register in one's own computer, and 'the same message' appears a microsecond later in one's friend's computer as a similarly structured string of switch positions in a register in the second machine. Unlike the letter written on paper no molecules have passed between them. The second computer has encoded the same structure of 0's and 1's as were encoded in the first computer. The message has been replicated by the replication of the patterns in the target register.

In short, the primary unit of organic evolution is the gene, a structure realizable in an indefinite number of molecular copies.

THE BIOMORPH WORLD

How could the transformation that can be achieved by the inheritance of minute variations in a genetic code be illustrated in a powerful and convincing way? Suppose one could create artificial beings, something like organisms but very simple, and control the choice of those that would realize some desirable characteristic. This characteristic would become more and more visible by the accumulation of minute changes generation after generation.

Dawkins's biomorphs are the products on the computer screen of the realization of a function with 9 variables. A sequence begins with a being of some primitive form generated by giving numerical values to each of the 9 variables. Just as the DNA strands realize the genes of a certain species of beings, a biomorph genome is a set of 9 numerical values assigned to the 9 variables in the generating program. Mutations can be modelled by changing the numerical value of one or more of the variables related in the relevant function. By randomly changing these values and selecting for the same desirable structural feature at each step, images with remarkably visually interesting structures can be brought into being.

The 'apparatus' with which the evolutionary experiment is conducted consists of the computer, its screen or VDU, the program and the human being who chooses which biomorph to 'breed' from in the next generation. 'The human eye has an active role to play in the story. It is the selecting agent. . . . The human tells the computer which one of the current litter of progeny to breed from.'[10] Here is an evolutionary lineage, with all but the chosen breeding biomorph represented (Figure 17).

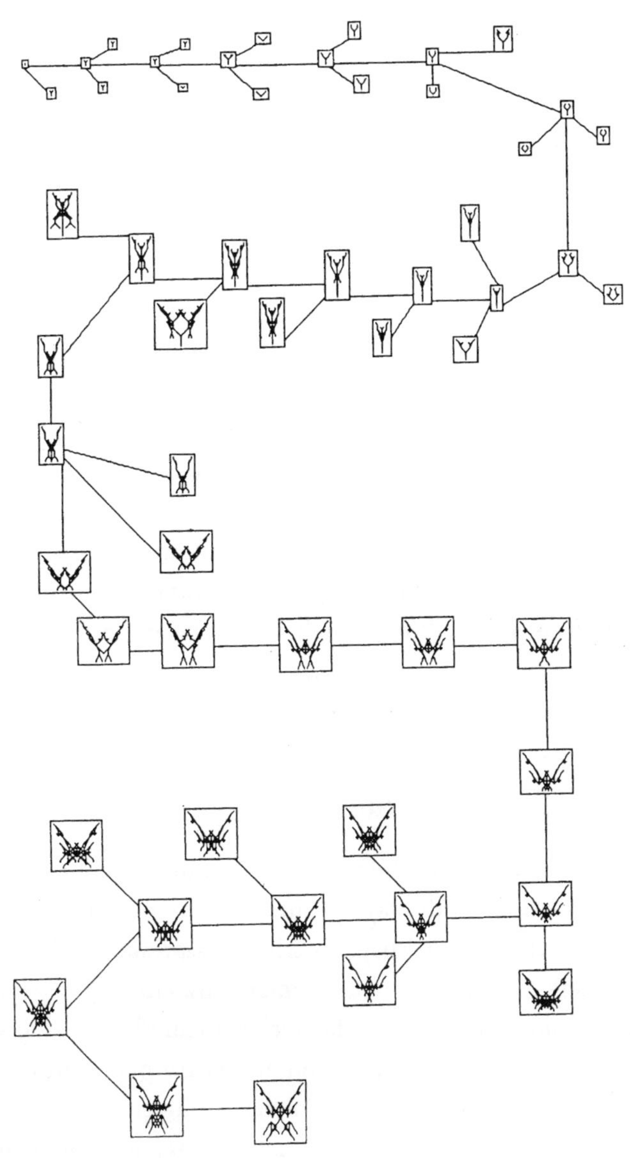

Figure 17. Lineage of biomorphs.

Partly as a result of challenges from Dawkins himself the 'game' of Darwinian evolution in Biomorphland has spread across the Internet. One of the most interesting has been developed by Alain Gogniat. Dawkins's original biomorphs reproduced asexually, much as aphids do for many generations. Gogniat's program displays sexual reproduction in that each new generation of biomorphs to be subject to selection is created from a combination of the 'genomes' of each of two 'parents'. We have not only random mutation but also the effect of recombination of genetic material from each of the two parents, just as it happens in sexual reproduction in the natural world.[11]

There have been several other interesting developments of the 'biomorph game' to be found on the Web. One of these can be singled out as coming closer to the real world than just an enlightening analogy. This is the 'hordeomorph' model. It has been used as a full-scale research tool. The project involved a program derived from Dawkins's biomorphs to simulate the genetic determination of the form of the ears of barley.[12]

Commentary

Galileo, Newton, Schrödinger, and Dawkins have all used imaginary situations to display the intelligibility of a new scientific idea or principle and perhaps to criticize a presupposition of an existing way of explaining a phenomenon by showing that it leads to an absurdity. Galileo used the horseman and his spear to show that it makes sense to treat vertical and horizontal motion independently, even though they are modes of movement of the same thing. Newton used his rotating globes to try to show that the idea of absolute space made sense. Later Mach objected that Newton had presupposed that the globes would still have mass in an empty world. Without that gratuitous presupposition there would be no tension in his imaginary string. Schrödinger used his imaginary cat in an imaginary box to show that a certain way of thinking about the deepest aspects of the physical world was unintelligible. A cat just cannot be in a superposition of the incompatible states 'alive' and 'dead'. So an electron cannot be in a superposition of states 'spin up' and 'spin down' before the experimental testing of which state *this* electron is in. Schrödinger favoured the idea that subatomic particles like electrons are really products of the experimental setup, and not extracted from a pre-existing state of the world. Dawkins used his biomorphs to show that the accumulation of minute variations in the genome could lead to very great changes in the organisms produced in accordance with the genetic 'instructions'. Unless one pays attention to the details of the evolutionary process in which minute variations in the genome are favoured or disfavoured in the process by which Interactors are reproduced, one might easily come to imagine that the fit of organ and environment was the work of a designer. Of course, the biomorph world is meant as an illustration of a principle rather than an explanation of the actual process of organic evolution. It is analogous to the breeding procedures on which Darwin modelled his conception of natural selection through a pattern of subtle analogies.

Imaginary beings do have an important part to play in the sciences. Some have been animals and plants, some have been gadgets that we can imagine as made of metal, wood, and string. The recruiting of computers and their powers to the constructing of models of living beings suggests at least one way in which living beings might be the core of experimental apparatus. The whole of a research programme might be implemented entirely in virtual worlds. The apparatus and instruments as well as the aspects of the world that these devices are created to detect, measure, or model might be simulated in computer programs. Looking back to the categories of apparatus illustrated in Chapters 7 and 8, Schrödinger's 'cat in a box' and Dawkins's biomorphs are logically speaking analogues. In the end that is how their plausibility and utility will have to be judged. We can be sure that a great many more such imaginary beings will be created to enrich the instrumentaria of the sciences.

Reflections: Sites for Moral Debates

The moral issues that have surfaced in recent years in popular and political debate about the uses of animals for scientific purposes have often been concerned with the application of the results of that research to practical matters. Should the planting and harvesting of genetically modified food plants be encouraged? Should animals be used to test medicines? Should soldiers have been exposed to the radiation from atomic explosions? And so on through a myriad of passionately debated issues. Do any of these debates touch on the topics of this book—the uses of animals and plants as instruments and apparatus? It can hardly be denied that they do.

Biology and psychology are sciences of living things, in both of which there has been a great deal of experimental research. Though experimental biology and more recently experimental psychology take living beings as their research materials, does it follow that the instrumentarium assembled for this purpose must also be organic? Biophysicists and biochemists use the instrumentarium of the relevant physical sciences in their researches. If we are concerned to track the way that moral issues arise in the biological and psychological sciences, the nature of the instrumentarium and the nature of the experimental material are not necessarily intimately related. Are the same moral principles relevant to the use of living beings as instruments and apparatus germane to the assessment of the standing of the beings which are the subjects of experiments? This book is about the instrumentarium, not about the beings on which it is used.

However, in many cases, instruments and apparatus are not independent of the research objects to which they are directed. Many of the experimental devices described in this book were organic, the living bodies of plants and animals. Then so were the subjects of the research. Hales's horses, Pavlov's dogs, Harlow's monkeys, and Milgram's 'teachers' were as much pieces of apparatus as they were topics of the research. The projects in which they were used were tests of putative biological and psychological hypotheses. The moral issues in such cases are more complex than those in which the origin of the items in the instrumentarium and the nature of the subjects of research can be discussed independently. How could one possibly do a study of the psychological conditions that underlay the carrying out of genocide unless people as possible perpetrators were not only subjects but constituents of the apparatus as well?

There are two aspects to the ethics of scientific research. One concerns the morality of the scientific community, a morality built around the idea of truth-telling and mutual trust, no matter what the topic nor the methods by which it is researched. We have seen the power of this aspect of the scientific moral order in the story of Kammerer and his toads. The other is the implicit principle that the ultimate ground of the morality of scientific research is the growth of knowledge. Whatever moral considerations emerge from reflection on experimental methods they have often been trumped by the over-riding principle that favours the growth of knowledge over all other goods. Suppose we couple this with the curiously arrogant piety of the predominantly Protestant scientists of the seventeenth century. They set aside all other considerations in their zeal to reveal the Universe as God's Creation. Keeping this in mind we have some inkling as to why few voices were raised against Robert Boyle and his circle, nor later against the good pastor Stephen Hales.

Whether the unfettered growth of knowledge, considered independently of its possible uses, is a moral absolute is of course highly debatable. It is a different debate from that concerning the uses of organic materials for the testing of medicines, for assessing the effects of consumer products on customers, or for other commercial purposes. These uses of organic beings have been set firmly outside the remit of my studies. Yet, the line I have drawn between experimental science and its practical applications could be argued to be much fuzzier than I have been making it out to be. When does a trial of the efficacy of a new drug cease to be only of medical interest and come to bear upon the scientific background that is also, in a way, under test at the same time? Does the kind of humane defence of drug trials spread its beneficent rays upon the parallel procedures of biochemical experimentation? On the other hand, if Milgram's experiment is supposed to throw light on the psychology of the perpetrators of the Holocaust must it not replicate at least some of the suffering of the victims?

Debating positions on the rights and wrongs of using animals and plants as instrument and apparatus whatever the subject of research seems to fall somewhere along three polarities.

One of these dimensions expresses attitudes to the accumulation of scientific knowledge. There are those who hold such accumulation to be an absolute good, to which all other considerations—moral, practical, and economic—must be subordinated. Others qualify this stand by reference to considerations of practical benefit and whether the resources drawn for the research might not be more beneficial to people if expended on other projects. There are also those who believe that the accumulation of scientific knowledge is a worthless enterprise anyway. Religious quietists such as traditional Buddhists do not do scientific research.

Another dimension reflects the degree to which we think that living beings are morally protected, that is, have inalienable rights to flourishing lives. This stance might be absolute, and would include plants as well as animals. A spindly, yellow plant in a dried out pot could be as morally offensive to someone as a gaunt and starving dog penned in behind a fence. Yet, if we base our intuitions on the good of human beings it is not hard to slip into thinking that some lesser beings could be sacrificed for the good of the lords of creation. In some places the death penalty for human beings is justified on the grounds that it is an effective deterrent to violence and morally defensible in terms of the good of all. There are those who take no thought for the rights of any living beings but themselves.

A third dimension reflects our willingness to ascribe mental lives to other organisms, and, in particular, to ascribe subjectivities more or less like our own. It is pretty hard to deny thoughts and feelings to a chimpanzee. The great number of dogs 'sacrificed' in some of the research programmes described herein suggests that this intuition is qualified for the treatment of 'man's best friend' by some very distinguished scientists. Cultivating bacteria and boiling up lichens do not invoke these intuitions at all.

To debate the moral issues that the uses of animals and plants as scientific equipment raises one must stand somewhere along these dimensions. No doubt there are other concepts relevant too.

The purpose of this book is to describe as clearly as possible the great variety of ways that living beings have been caught up in the projects of experimental science—not simply as topics to be investigated but as in whole or part the apparatus and instruments by which research of all kinds has been accomplished. I leave the working out of moral arguments to those more qualified than I am to reach just and ethically sound conclusions. I offer these guide lines only as a starting point for readers to reflect on the ethical aspects of the scientific practices I have described.

End Notes

Chapter 1: Instruments and Apparatus

1. Matt Ridley, *Nature via Nurture* (London: Fourth Estate, 2003).
2. Stephen Hawking, *A Short History of Time* (London: Bantam, 1988).
3. I take this useful expression from Robert Ackerman's *Data, Instruments, and Theory* (Princeton, NJ: Princeton University Press, 1985).
4. R. E. Kohler, *Lords of the Fly* (Chicago/London: Chicago University Press, 1994), 115.
5. These two terms are the current favoured euphemisms for telling us that the animal subjects died or were killed in the course of the experiment.

Chapter 2: Detecting

1. J. Palmer, *Rev. Abraham Bennett 1749–1799* (2003). Available at: http://www.wirksworth.org.uk.
2. L. Galvani, *De Viribus Electricitatis* [*Commentary on the Effects of Electricity on Muscular Motion*], trans. M. G. Foley (1791; Norwalk, CT: Burndy Library, 1953).

3. As noted by I. B. Cohen in his introduction to the 1953 English translation of Galvani's treatise (*De Viribus Electricitatis*).
4. As noted by I. B. Cohen in his introduction to the 1953 English translation of Galvani's treatise (*De Viribus Electricitatis*) 45.
5. B. Franklin, *Experiments and Observations on Electricity*, ed. I. B. Cohen (1774; Cambridge, MA: Harvard University Press, 1941), 256–7.
6. D. H. S. Richardson, *The Vanishing Lichens* (Newton Abbott: David and Charles, 1975), 16.
7. Ibid. 83.
8. Source: *BBC News*, 30 December 1986.
9. I am grateful to Dr Harry Charlton and Professor David Sherratt for generously providing me with tutorials on transgenic techniques and processes.
10. M. A. Liebert, 'Investigator Profile: Interview with Zhiyuan Gong', *Zebrafish* **1**(3) (2004), 185–9.
11. Z. Zeng, T. Shan, Y. Tong et al., 'Development of Oestrogen Responsive Transgenic Medaka for Environmental Monitoring of Endocrine Disruptors', *Environmental Science and Technology* **39** (2005), 9001–8.
12. Z. Gong, H. Wan, T. L. Tay et al., 'Development of Transgenic Fish for Ornamental and Bioreactor [Use] by Strong Expression of Fluorescent Proteins in the Skeletal Muscle', *Biochemical and Biophysical Research Communications* **388** (2003), 58–63.
13. Ibid.
14. *New Scientist* (24 November 2003).
15. This passage is taken from the Authorized Version, Genesis 7:17–20. The spelling has not been modernized.
16. A fascinating biographical pamphlet has been published by the Oxford Museum of the History of Science. It includes the extraordinary story of Buckland eating the preserved heart of one of the Kings of France.
17. This was the usual spelling in the nineteenth century.
18. Quoted in N. A. Rupke, *The Great Chain of History: William Buckland and the English School of Geology (1814–1849)* (Oxford: Clarendon Press, 1983), 33.

19. The story of the Montgolfiers and their balloons is brilliantly told by C. C. Gillispie, *The Montgolfier Brothers and the Invention of Aviation: 1783–1784* (Princeton, NJ: Princeton University Press, 1983).
20. Ibid.
21. Ibid. 40.
22. For the account that follows, though much has been written on the topic, I have drawn heavily on the comprehensive account of the Russian space dogs in *Space Today Online/Dogs in Space*, see http://www.spacetoday.org/Astronauts/Animals/Dogs.html.

Chapter 3: Measuring

1. R. Tavernor, *Smoot's Ear: The Measure of Humanity* (New Haven, CT: Yale University Press, 2007).
2. Unfortunately, as Tavernor notes, there are seven of these feet to the presumed total double yard spread. I have personally verified Tavernor's observation, taking care not to place my hands on the exhibit.
3. An excellent explanatory catalogue of traditional English measuring units is due to Russ Rowlett and can be found at http://www.unc.edu/~rowlett/units.
4. J. D. North, *God's Clockmaker: Richard of Wallingford and the Invention of Time* (London: Hambledon and London, 2005).
5. G. Galileo, *Dialogues Concerning Two New Sciences* (1665; New York: Dover, 1914), 179.
6. J. Phillips, *Memoirs of William Smith LLD* (London: John Murray, 1844).
7. *Washington Post* (26 June 2004), B06.
8. T. Gold, *The Deep Hot Biosphere* (New York: Springer Verlag, 1999). The book makes fascinating reading with its calm and measured exposition of some very remarkable ideas.
9. We will return to the nature and classification of simple organisms in more detail in Chapter 8.
10. J. R. Parkes, G. Webster, B. A. Cragg et al., 'Deep Sub-seafloor Prokaryotes Stimulated at Interfaces over Geological Time', *Nature* **436** (2005), 390.

11. R. Summons, L. L. Jancke, J. M. Hope et al., '2-Methylhopanoids as Biomarkers for Cyanobacterial Oxygen Synthesis', *Nature* **400** (1999), 554–7.
12. M. Rosing, 'C-Depleted Carbon Microparticles in >3,700-M in Seafloor Sedimentary Rocks from West Greenland', *Science* **283** (1999), 674. Rosing's result has not been accepted by some geologists and remains controversial.
13. Thanks to Stephen Moorbath for helpful comments on the dating of the earliest organisms.
14. I am grateful to Ronnie Liljegren for a personal memoir of this remarkable man. See also R. A. Manten, 'Lennart von Post and the Foundation of Modern Palyology', *Review of Palaeobotany and Palyology* **1** (1967), 11–22.
15. H. A. Hyde and R. F. Adams, *An Atlas of Airborne Pollen Grains* (New York: Macmillan, 1958).
16. G. Erdtmann, *An Introduction to Pollen Analysis* (Waltham, MA: Chronica Botanica, 1943).
17. '500,000 Years of Climate History Stored Year by Year' (14 March 2007), available at http://www.physorg.com/news93087836.html.
18. D. H. S. Richardson, *The Vanishing Lichens* (Newton Abbot: David and Charles, 1975), 45.
19. S. J. Brooks, 'Fossil Midges as Palaeoclimatic Indicators for the Eurasian Region', *Quaternary Science Review* **25**(15–16) (2006), 1894–1910.
20. Of course, at least some of this carbon is of organic origin, so even radiocarbon dating has a link to once living beings.

Chapter 4: Extending an Established Domain

1. Finding my way around this voluminous correspondence was greatly facilitated by the work of R. G. Frank, *Harvey and the Oxford Physiologists* (Berkeley: University of California Press, 1980).
2. William Harvey, *De Motu Cordis*, trans. C. D. Leake (1628; Springfield, IL: C. C. Thomas, 1978), 104.
3. 'Boyle, Hon Robert', in *Dictionary of National Biography* (Oxford: Oxford University Press, 2004), v.100–8.
4. 'Wren, Christopher', ibid. lx.406–19.

5. W. C. Gibson, 'The Biochemical Pursuits of Christopher Wren', *Medical History* **14** (1970), 331–41.
6. R. Lower, *Tractatus de Corde* (1669), trans. K. J. Franklin, in *Early Science in Oxford*, vol. 9, ed. P. K. Gunther (Oxford: Oxford University Press, 1932), 1–120.
7. Robert Boyle, *The Works of the Hon. Robert Boyle* (London: Millar, 1744), vi.450.
8. Frank, *Harvey and the Oxford Physiologists*, 176–7.
9. Stephen E. Hales, *Statical Essays*, vol. 1, *Vegetable Statics*; vol. 2, *Haemastatics*, 3rd edn (London, 1769). All descriptions of these experiments by Hales are taken from these volumes. *Vegetable Statics* first appeared in 1727, followed by *Haemastatics* in 1733.
10. Hales, *Haemastatics*, 2.
11. Ibid. 66.
12. Hales, *Vegetable Statics*, Exp. II.
13. Hales, *Vegetable Statics*, Exp. VII.
14. Y. L. Shevchenko and J. E. Tsitlik, 'The 90th Anniversary of the Development of N. S. Korotkoff of the Auscultatory Method Of Measuring Blood Pressure', *Circulation* **94**(2) (1996), 116–18.
15. J. Mayow, *Medico-Physical Works, Being a Translation of Tractatus Quinque Medico-Physici* (1674), trans. A.C.B. and L.D. (Edinburgh: The Alembic Club, 1924), 84. Subsequent page references in text.
16. Hales, *Statical Essays*, 236. Subsequent page references in text.

Chapter 5: Exploring a New Domain

1. T. Cao, J. Vacanti, K. T. Paige et al., 'Transplantation of Chondrocytes Utilizing a Polymer-Cell Construct to Produce a Tissue-Engineered Cartilage in the Shape of a Human Ear', *Plastic and Reconstructive Surgery* **100** (2) (1997), 297–302.
2. A. R. Gardner and B. T. Gardner, 'Teaching Sign Language to a Chimpanzee', *Science* **45** (1969), 664–72.
3. Ibid.
4. H. S. Terrace, *Nim* (New York: Columbia University Press, 1987), 184.
5. S. Asch, *Social Psychology* (Oxford: Oxford University Press, 1952).

6. T. Blass, *The Man who Shocked the World* (New York: Basic Books, 2004).
7. S. L. Milgram, *Obedience to Authority* (New York: Harper Torchbooks, 1983), 21. Subsequent page references in text.
8. P. Zimbardo, *The Lucifer Effect* (London: Rider, 2007).
9. E. Goffman, *The Presentation of Self in Everyday Life* (Garden City, NY: Doubleday, 1959).
10. C. L. Sheridan and R. G. King, 'Obedience to Authority with an Intended Victim', *Proceedings of the American Psychological Association* 2 (1972), 165–6.

Chapter 6: Testing Hypotheses

1. I. Pavlov, *The Working of the Digestive Glands*, trans. W. H. Thompson (1897; Birmingham, AL: Classics of Medicine, 1980), 15.
2. W. Beaumont, *Experiments and Observations on the Gastric Juice and the Physiology of Digestion* (Plattsburgh, NY: Allen, 1833).
3. B. P. Babkin, *Pavlov: A Biography* (London: Gollancz, 1951).
4. I. Pavlov, *Lectures on the Work of the Digestive Glands*, trans. W. H. Thompson (1903; London: Charles Griffin, 1980), 53.
5. W. M. Bayliss and E. H. J. Starling, 'The Mechanism of Pancreatic Secretion', *Journal of Physiology* **28** (1902) 325–53.
6. J. Bowlby, *Attachment and Loss*, 2 vols (New York: Basic Books, 1969).
7. D. Blum, *Love at Goon Park: Harry Harlow and the Science of Affection* (Cambridge, MA: Perseus Books, 2002).
8. H. F. Harlow, *Learning to Love* (San Francisco: Albion, 1971).

9. It is worth commenting that on the very same campus as Harlow's monkey lab Eileen Walzer was pursing her analysis of romantic love in terms of cost–benefit analyses in which the partners continually monitored their relationship for profit and loss!

10. Harlow, *Learning to Love*, 18–19.
11. J. Goodall, *In the Shadow of Man* (London: Weidenfeld and Nicholson, 1989).

12. D. Sumilo, L. Asokliene, A. Bormane et al., 'Climate Change Cannot Explain the Upsurge of Tick-Borne Encephalitis in the Baltics', *PLoS ONE* **2**(6) (2007), e500.

Chapter 7: Modelling Individuals

1. L. K. Altmann, *Who Goes First? The Story of Self-Experimentation* (Berkeley: University of California Press, 1998).
2. C. Barnard and C. B. Pepper, *Christiaan Barnard: One Life* (London: Harrap, 1970), 332.
3. Ibid.
4. Ibid. 336.
5. R. R. Lower and N. E. Shumway, 'Studies on Orthotopic Homotransplantation of the Canine Heart', *Surgical Forum* **11** (1960), 18–19.
6. *USA Today* (10 February 2006).
7. P. Zimbardo, *The Lucifer Effect* (New York: Random House, 2007).
8. This example is from *The Times* (9 July 2004).
9. X. Zhang, J. M. Beaulieu, T. D. Sotnikova et al., 'Tryptophan Hydroxylase-2 Controls Brain Serotonin Synthesis', *Science* **305** (2004), 217.
10. L. Cervo, A. Canetta, E. Calcagno et al., 'Genotype-Dependent Activity of Tryptophan Hydroxylase-2 Determines the Response to Citalopram in a Mouse Model of Depression', *Journal of Neuroscience* **25** (2005), 8165–72.
11. A. Van Den Bogaert, K. Sleegers, S. De Zutter et al., 'Association with Brain-Specific Tryptophan Hydroxylase, Tph, with Unipolar and Bipolar Disorder in a Northern Swedish Isolated Population', *Archives of General Psychiatry* **63** (2006), 1103–10.
12. A. Caspi, J. McClay, T. Moffitt et al., 'Role of Genotype in the Cycle of Violence in Maltreated Children', *Science* **297** (2002), 851–4.
13. P. Marsh, W. Rosser, and R. Harré, *The Rules of Disorder* (London: Routledge, 1977).
14. *The Times* (9 July 2004).
15. B. J. Marshall, *Helicobacter Pioneers* (Oxford: Blackwell, 2002), 195.

16. Ibid. 196.
17. J. Eugenides, *Middlesex* (New York: Farrar, Straus, Giroux, 2002), 478–9.

Chapter 8: Modelling Worlds

1. I am grateful to John Clarke for his help in preparing this section. He confessed that the fact that the Olivia de Havill and movie, *The Snake Pit*, was showing at the local cinema may have prompted the choice of name.
2. D. H. Chitty, *Do Lemmings Commit Suicide?* (Oxford: Oxford University Press, 1996).
3. J. R. Clarke, 'Influence of Numbers on Reproduction and Survival in Two Experimental Vole Populations', *Proceedings of the Royal Society B* **144** (1955), 68–85.
4. J. R. Clarke, 'The Aggressive Behaviour of the Vole', *Behaviour* **9** (1956), 1–23; *idem.* 'The Effect of Fighting on the Adrenals, Thymus and Spleen of the Vole (*Microtus agrestis*)', *Journal of Endocrinlogy* **9** (1953), 114–26.
5. Clarke, 'Aggressive Behaviour of the Vole', 18.
6. R. S. Root-Bernstein, 'Mendel; and Methodology', *History of Science* **27** (1983), 275–95.
7. C. Singer, *A Short History of Science* (Oxford: Clarendon Press, 1959), 505.
8. A. Smith, *Blind White Fish of Persia* (Harmondsworth: Penguin, 1990).
9. T. G. Dobzhansky, *Genetics of the Evolutionary Process* (New York/London: Columbia University Press, 1970).
10. R. E. Kohler, *Lords of the Fly: Drosophila Genetics and the Experimental Life* (Chicago: Chicago University Press, 1994), 123.
11. J. Weiner, *The Beak of the Finch* (London: Cape, 1995).
12. T. B. Brock, *Thermophilic Organisms and Life at High Temperatures* (New York: Springer-Verlag, 1978), 303–36.
13. S. Henahan, 'Going to Extremes', *AccessExcellence* (1997), 1–2. Available at http://www.accessexcellence.org/WN/SUA10/extreme797.php.

Chapter 9: Practising Deception

1. J. Fyfe, *Lysenko is Right* (London: Lawrence and Wishart, 1950), 38.
2. D. Joravsky, *The Lysenko Affair* (Cambridge, MA: Harvard University Press, 1970).
3. V. N. Soyfer, *Lysenko and the Tragedy of Russian Science*, trans. L. and R. Gruliow (New Brunswick, NJ: Rutgers University Press, 1994), 35.
4. Ibid.
5. M. Gardner, *Fads and Fallacies in the Name of Science* (New York: Dover Books, 1957).
6. Soyfer, *Lysenko and the Tragedy of Russian Science*, 89.
7. G. Holton, *Thematic Origins of Scientific Thought* (Cambridge, MA: Harvard University Press, 1988), 453.
8. A. Koestler, *The Case of the Midwife Toad* (New York: Random House, 1971).
9. G. K. Noble, 'Kammerer's *alytes*', *Nature* **118** (1926), 209–10.
10. The material for this section is drawn almost exclusively from Koestler's *Case of the Midwife Toad*, particularly Chap. 1, Sect. 4. Koestler himself credits the details to Kammerer's daughter, Lacerta.
11. F. Close, *Too Hot to Handle: The Race for Cold Fusion* (New York: Penguin, 1992), 9.
12. For a detailed story of the social, commercial, political, and personal aspects of this episode see E. F. Mallove, *Fire from Ice* (New York: Wiley, 1991).

Chapter 10: Inventing Novel Beings

1. Erwin Schrödinger, 'Autobiographical Sketches', in *What is Life?* with *Mind over Matter* and *Autobiographical Sketches* (Cambridge: Cambridge University Press, 1992), 184.
2. Schilpp, P. A. (1970) *Albert Einstein: Philosopher-Scientist.* London: Cambridge University Press.
3. E. Schrödinger, 'The Present Situation in Quantum Mechanics', in J. A. Wheeler and W. H. Zorak (eds), *Quantum Mechanics and Measurement* (Princeton, NJ: Princeton University Press, 1983).

4. K. Fine, *The Shaky Game: Einstein, Realism and the Quantum Theory* (Chicago: Chicago University Press, 1986), 78–84. Fine credits Einstein with a story that bears on the same point. In Einstein's story there is an explosive hidden in a box with a random trigger.
5. Schrödinger, 'Present Situation in Quantum Mechanics', 157.
6. Schrödinger, 'Present Situation in Quantum Mechanics'.
7. For an admirably clear and more detailed discussion of the meaning of Schrödinger's Cat Paradox see M. Lockwood, 'What Schrödinger Should Have Learned from His Cat', in M. Bitbol and O. Darrigol (eds), *Erwin Schrödinger* (Paris: Editions Frontieres, 1992), 363–82.
8. R. Dawkins, *The Blind Watchmaker* (New York/London: Norton, 1986).
9. R. Dawkins, *The Selfish Gene* (Oxford: Oxford University Press, 1976).
10. Dawkins, *The Blind Watchmaker*, 57.
11. The program can be found at Gogniat's website http://www.cs.bgu.ac.il/~sipper/biomorphs/evolution.html.
12. G. H. Buck-Sorlin, O. Kniemeyer, and W. Kurth, 'Integrated Grammar Representation of Genes, Metabolites and Morphology: The example of hordeomorphs', in C. Godin (ed.), *4th International Workshop on Functional-Structural Plant Models* (2004), 366–89.

Index

Acquired characteristics, inheritance of, 244, 245, 251, 258
Aggression and glandular development, 214
Air, composition, 118
Altitude dangers, animal tests for, 50
Analogy
  of behaviour, 13
  of process, 13–14
Animal hydraulics and pneumatics, 91
Animal spirits, 23
Animals as detectors, non-scientific uses, 56
Apparatus
  and research subjects, 298
  as abstraction, 6–7, 12
  as analogue models, 6, 7–8, 11–12
  as domestic version, 12
Aristotelian 'causes', 92
Artificial organs, 123, 124–127
  basic technique, 127
Baboon liver as dialysis machine, 179–180
Bacteria, aerobic and anaerobic, 71
Balloon design, 49
Biological systems, role of blood, sap and air, 89
Biomorphs, 271
Biopolymers, uses of, 127
Biosphere, deep and hot, 74–75
Black body radiation, 275
Blood circulation,
  parameters, 92–93, 107, 147
  problems, 96
Blood motion in the lungs, 108
Blood pressure,
  and volume, 109–110
  measurement, 108, 116–118
  relationships, 107
  systolic and diastolic, 109
Blood transfusions, 94, 100
  methods, 100–101

Blood transfusions (*cont.*)
  pressure differences, 102
Breathing experiments, 119–120

Calciferous stones, 106
canaries in mines, 35–36
Carbon dating, organic features, 83
Cat paradox, 284–285
Celsius scale, origins, 64
Chironomids and temperature, 84
Chromosomes, role in genetics, 222
Clarke, John, career, 204–206
Clepsydra and Galileo's pulse, 66–67
Climate change,
  dating, 76–82
  detection, 40–47
Cold fusion, 265–267, 268
Computer generation of images, 287, 292
Corpuscularian theories, 94–95
Corpuscularian universe, 274
Crucial experiment, weak logic of, 148
Cruelty, its origins, 123
Cumulative selection, 288

Daphne Major,
  as natural experiment, 234–235
  as world model, 231, 236
Detectors, need for, 20–21
Digestion,
  control mechanisms, 151
  humoric (chemical) control, 156–158, 174
  nervic control, 151, 154, 174
Diluvianism, 42–43, 46–47
Dobzhansky, Theodosius, career, 226–227, 230–231
Dogs and horses as apparatus, 90, 94
Dogs as detectors, 53–55
Drosophila melangaster, 219 ff

Earth, early conditions models, 236–7
Elasticity of air, 119, 121
Electricity,
  from dissimilar metals, 29, 30
  types of, 23, 26, 28
Electron
  as particle, 276
  as wave, 276
Electron model of atom, 271
Electroscope, inorganic, 22
Encephalitis, epidemics, 172
Encephalitis,
  plethora of alternatives, 173–174, 175
  tick hypothesis, 172–173
Engineering and science, 124, 176–177
Environmental influences, Zimbardo experiment, 142–143
Ethical framework, 17, 18
Evolution, metaphors, 291
Experimental design, intensive v. extensive, 90, 138, 146
Experimental equipment,
  animals and plants as, 2, 3–4
  automatic, 2
  inorganic 2, 3
Extraterrestrial tests, 48–55
Extremeophiles, 237

Fahrenheit scale,
  duodecimal system, 63–64
  origins, 65
Fertility and aggression, 213
Field studies, observations, 224, 228

Filial love
comfort hypothesis, 160, 175
experimental methodology, 165–169, 170, 171
nutrition hypothesis, 160, 175
research background, 161
Finches,
observation program, 231–232
species of, 233–234
Fluorescent indicators, 40
Fossil types and strata, 69
Fossils as ordinal time scales, 67
Fossils, earliest, 75
Fraud,
kinds of, 243
motivations, 242
Frog's legs as electroscope, discovery, 25
sensitivity, 30

Galen, refutation of, 93
Gastric fistulas, 154
Gastric-juice factory, 152–153
Gene
expression, 186
insertion, 37–39
variation, effects of, 187–188
Genes
and behaviour, 184–193
replication, 289
Genetic explanations of human behaviour, 201
Genetic processes as mechanism of Darwinian evolution, 227–228
Genetics of peas, 218–219
Geological maps, 68–69
Glacial effects, 44
Greenland climate change, 87
Growth as cell division, 216–217
Growth of knowledge, 299

Hales, Stephen, career, 103–104
Harlow, Harry, career, 163–165
Heart
surgery, 182
transplantation, 183–184
volume of, 108
Helicobacter pylori,
observed, 195–1
the puzzle, 193, 194–195
Hidden variables, lack of, 278
High and low population density model worlds, 207–208
Homologous heart experiment, 182–183
Hopanoids as molecular signatures, 73–74
Human aggression,
childhood maltreatment, 192
genetic evidence, 191
Human bodies and space measurement, 60–62
Human cruelty, hypotheses, 138
Hybrid apparatus based on dog, 97
Hydrocarbons,
biogenic v. abiogenic origins, 72, 73
surface origin v. depth origin, 74, 75
Hyena's den, 42, 45–47

Ice worms, 238–240
Igneo-aërial spirit, 119
Index fossils, 77
Indicators, 31–33
Injection, methods for, 97, 98–99

Instrumentarium,
contents, 8–9
detectors, 5
measuring devices, 5–6
mode of working, 10
Intensive design, 106, 116, 193

Kammerer, P., character, 256, 261
Koch's postulates, 194, 197
Korotkoff sounds, 116

Laboratory colony as model of natural population, 230
Laboratory populations, experiments, 220, 227, 229
Lake bed deposits as dating devices, 77, 81–82
Lamarckian theory, tests of, 225
Language acquisition, chimpanzee, 130–131, 132–133
Language of apes, 123
Language studies, limits of apparatus, 134
Language,
rules of use, 131
species specificity, 130, 132
syntax and semantics, 131
Leyden jar theory, 26, 27, 28
Lichenometry,
calibration, 82
principle of, 82
Lichens,
source of dyes, 32
structure, 32
Litmus,
extraction, 33
uses of, 31
Living cell, discovery, 215–216
Lysenko, T.,
character, 245
political career, 250, 251, 252

Majority influence, 139
Manned flight, 52
Measurement,
fixing of units, 57, 59
reliable devices, 57, 59
Mechanical clocks, 65
Mendel, Gregor, career, 217, 219
Metrical precision, 103, 104, 105
Metrication, 62
Mice as models, 178, 184–193
Midge thermometers, 83–85
Milgrim experiment,
analysis of results, 141, 145
animal component, 143–145
human components, 139
procedure, 140–141
Milkweed, 35
Millikan's oil-drop experiment, 254–255
Model worlds,
in process, 210–212
significance, 203, 240–241
Modelling, double analogy, 186, 187
Molecular fossils, 71
Mollusc
distribution and temperature, 86–87
thermometers, 86–87
Monkey mother surrogates, 170
Moral concerns, 18, 177–178
Moral debate, polarities, 299–301
Morality in science, varieties, 298
Morality of science v. politics, 245
Mouse,
depression, 188
role in pinna experiment, 128

Mouse-human aggression analogy, 190
Mouse-human depression analogy, 189
Muscle swelling, 111–112

Nerves, structure, 23
Noah's Ark, 41–42

Ordinal time measurements, 58
Organisms in science, 14–15
Oxford 'cognoscenti', 90, 94

Palynology, 78
Particles as local manifestations, 281
Pavlov pouch, 154
Pavlov, Ivan, career, 149–152
Peat bogs as ordinal records, 78
pH values, 31
Physiological surgery, 149, 157, 150–151, 157, 159
Pilot plants, dogs as, 178, 181–184
Pinna cartilaginous frame, 128
Plasmids, 39
Pollen
- deposition, 77, 81
- grains and plant species, 79

Popular science, 2
Population cycles, 205–206
Positivism in physics, 277
Principle of Faunal Successions, 69
Prokaryotes, bacteria and archaea, 74

Quantum mechanics, problems of modelling, 270–271
Quarries and canals as sources of fossils, 68

Radiation and matter, 275
Replicators and Interactors, 286, 290
Returning stroke, 27
Reuptake inhibitors, 189
Russian space programme, 52

Safety lamp, 36
Sap,
- analogy to blood, 112–113
- mechanism of motion, 114
- transpiration through leaves, 113–114
- velocity in roots and stem, 113

Scientific realism, 271, 277
Self-experimentation, 178
Self-infection experiment, 198–199
Serotonin,
- high levels, 189
- low levels, 187

Sex linked conditions, 223
Sexual reproduction modelled, 294
Smoot's ear, 60
Space flight dangers, animal tests. 53–55
Space, standard units, 60–62
Spin, up and down, 280
Superposition of states, 282–283
Syntactical skills, Terrace's analysis, 135–137

Temperature, units, 63–64
Thermophiles, as models of early life forms, 237–238
Thought experiments, 269–270, 294–296
Time measurement,
- devices, 64–65
- standard units, 58, 59

Titan as methane atmosphere model, 238–239

Toads,
breeding types, 257–258
experiments, 257, 258–259, 262, 265, 268
fraud, 260, 263, 264
key specimen, 259, 260
Toxicogenomics, 39
Transgenic fish, 36–40
Tropical animal remains, 46
Type and token, 289

Umwelt, 204

Vacanti mouse, misunderstandings, 129
Vagus nerve, function, 155
Ventilators, 104
Vernalization, 246
Virtual animals, 19
Vole behaviour, aggression/submission, 209

Washoe, life story, 137
Waves and particles, 279–280
Weather forecasting,
folk, 34
insect swarms, 34–35
Wheat crop,
experimental plots, 248
field trials, 247
fraudulent results, 249, 253
Wren, Christopher, career, 97–98

Index of Names

Adams, R., 78
Aristotle, 90, 91, 92
Arlande, D' M. 52

Barnard, C., 17, 179–180, 181, 183, 184
Bateson, W., 259, 260, 263
Bayliss, W., 156–159, 174, 175
Beaumont, C., 152
Bennett, A., 22
Berkowitz, L., 165
Berthollet, C.L. and A. B., 33
Bischel, R., 4, 82, 87
Bizozzero, G. 194–195
Blum, D. 163
Boag, P. and L., 234
Bogaert, van den, 189
Bohr, N., 278
Bondi, H., 72
Born, M., 279
Bowcott, O., 34
Bowlby, J., 160
Boyle, R. 91, 94–96, 99, 101, 121, 299
Brenner, S., 220
Brock, T. 237–238
Brooks, S. 85
Buckland, W., 42–47

Caron, M., 187, 192
Caspi, A., 189–192
Castle, W., 220
Celcius, A., 64
Cervo, L. 188
Cherverikov, S., 224
Chitty, D., 205, 206, 207
Chomsky, N., 130, 131, 132
Clarke, J., 9, 11, 12, 202, 203–214

D'Alibard, T-F, 28
Darwin, C., 11, 13, 232, 286
Dawkins., R.. 1, 272, 286–294, 296
Dirac, P., 282
Dobzhansky, T., 14, 202, 224, 226–231

Einstein, A., 275, 278–279

Fahrenheit, D. 57, 62–64
Fahrenheit, Frau, 59
Faraday, M., 3, 9
Fisher, C., 239
Fleischmann, M. 265–267, 268
Fontana, F., 23
Franklin, B., 28
Freud., S., 261
Fyfe, J., 245

Galeazzi, D., 25
Galeazzi, L. 25
Galen, 90, 93
Galileo Galilei, 66–67, 102, 269–270, 294
Galvani, L. 14, 22–30
Gardner, B and R., 10, 130–137, 146
Gargarin, Y., 53
Goffman, E., 141
Gogniat, A., 294
Gold, T., 71–75
Gong, Z. 39–40
Goodall, J., 166
Grant, P. and R., 1, 8, 13, 16, 202, 231–236, 252
Grew, N., 113

Hales, S., 7, 14, 16, 90, 91, 103–114, 120, 121, 147, 299
Harlow, H., 4, 8, 160–171, 172, 175, 223
Harvey, W. 91, 92–94, 99, 107, 147
Hawking, S., 2
Heisenberg, W., 277, 282
Henderson, M. 186
Hooke, R., 215–216
Horvitz, R., 220
Hoyle, F., 72
Hyde, H., 78

Illingworth, A., 35

Jones, G., 87
Joravsky, D., 248

Kammerer, P. 255–265, 267
King, R. 145–147
Kohler, R., 9
Korotkoff, N., 116
Kruschev, N., 252
Kudryavtsev, N. 71

Lamarck, J., 225, 259, 265
Langer, R., 127
Lavoisier, A., 6, 7, 49
Lavoisier, M.-A., 25
Linnaeus, C. 64
Litt, T. 81–82
Lor, de M., 28
Lorenz, K. 212
Louis XVI, 49, 52
Lower, R. (the American) 180–183
Lower, R.(the Cornishman) 91, 95, 96, 99, 100, 101–102
Ludwig, C., 150
Luria, A., 244
Lyell, C. 43, 76
Lysenko, D., 246
Lysenko, T., 244–252, 255, 266, 267

Mahler, A., 261, 268
Mahler, G., 261
Mahon, Lord, 27
Malpighi, M., 100, 113

Marie-Antionette, 49
Marshall, B., 15, 178, 193–200, 201
Mathur, K., 255
Maxwell, J., 275
Mayow, J., 118–120, 121
Mendel, G., 1, 12, 14, 217–219, 248
Michurin, I., 244
Milgrim, S., 9, 15, 138–145, 299
Mill, J. S., 241, 253
Millikan, R., 254–255, 267
Moffatt, T., 189–192
Montgolfier, E., 48, 49, 50
Montgolfier, J., 48, 49
Morgan, T., 221–223, 225, 227, 228
Morland, M., 44
Musso, H., 32

Napoleon Bonaparte, 30
Newton, I. 269, 294
Noah, 41
Noble, G., 260, 264
Nollet, J., 22

Ockham, William of, 253

Parkes, R. 74
Pavlov, I., 9, 10, 12, 14, 16, 54, 149–160, 172, 174, 215
Payne, F. 225–226
Planck, M., 275
Polyani, M., 206
Pons, S., 265–267, 268
Post, von L., 4, 77–78
Pribram, K., 137
Price, T., 235
Prizbram, H., 259, 262

Randolph, S., 172, 173, 175
Rennie, A., 68
Richard, of Wallingford, 65
Ridley, M., 2
Root-Bernstein, R., 219
Rosier, de P., 52, 53
Rosing, M., 75

Samoïlov, V., 153
Sapegin, A., 249
Scarborough, C. 97
Schrödinger, E., 270, 271, 273–285, 294, 296
Shaw, B., xii
Sheridan, C. 145–147
Shumway, N., 9, 181–184
Singer, C., 225
Skinner, B., 161, 164
Smith, W. 14, 67–70, 76
Smoot, O., 60
St Martin, A., 151
Stalin, J., 250
Starling, E., 156–159, 174, 175
Stone, C., 164
Stukely, W., 107
Sturtevant, J., 223
Sukachev, V., 252
Sulston, J., 220
Sumilo, D., 172
Summons, R., 75

Tajfel, H., 163
Tavernor, R., 60
Terman, L., 164, 165
Terrace, H., 135–136, 146
Theodoric of Freiberg, 7
Thomson, J. J., 276
Tinbergen, N., 212

Trybom, F., 77
Tsion, I., 150

Vacanti, C., (snr.) 127
Vacanti, C., (jnr.) 127, 128
Vacanti, J., 126–127
Valera, de R., 274
Vavilov, N., 250, 251, 252
Volta, A., 22, 30
Vygotsky, L. S., 244

Walter, B., 261
Warren, R., 178, 193–200, 201
Washkansky, L., 183
Watson, J., 164
Webb, E., 68
Weinberg, J., 165
Weismann, A., 258
Wiesenthal, G., 264
Wittgenstein, L., 213
Wren, C., 14, 91, 96, 97–99, 100

Zimbardo, P., 142